AF560384

CHEMISTRY OF ORGANIC NATURAL PRODUCT

CHEMISTRY OF ORGANIC NATURAL PRODUCT

Ulag Mahadevan

ANMOL PUBLICATIONS PVT. LTD.
NEW DELHI-110 002 (INDIA)

ANMOL PUBLICATIONS PVT. LTD.
Regd. Office: 4360/4, Ansari Road, Daryaganj,
New Delhi-110002 (India)
Tel.: 23278000, 23261597, 23286875, 23255577
Fax: 91-11-23280289
Email: anmolpub@gmail.com
Visit us at: www.anmolpublications.com

Branch Office: No. 1015, Ist Main Road, BSK IIIrd Stage
IIIrd Phase, IIIrd Block, Bangalore-560 085 (India)
Tel.: 080-41723429 • Fax: 080-26723604
Email: anmolpublicationsbangalore@gmail.com

Chemistry of Organic Natural Product

First Edition, 2011

ISBN 978-81-261-4594-2

PRINTED IN INDIA

Printed at Mehra Offset Press, Delhi.

Contents

Preface

This book reviews in a concise and manageable way the progress in all key areas of natural products chemistry since 1984. The most significant advances are highlighted over a wide field of chemistry, structure, synthesis and biosynthesis. This book provides a unique and superb entry into the vast literature on the subject. Comparisons of natural products from microorganisms, lower eukaryotes, animals, higher plants and marine organisms are now well documented. This book provides an easy-to-read overview of natural products. It includes twelve chapters covering most of the aspects of natural products chemistry.

Each chapter covers general introduction, nomenclature, occurrence, isolation, detection, structure elucidation both by degradation and spectroscopic techniques, biosynthesis, synthesis, biological activity and commercial applications, if any, of the compounds mentioned in each topic. The introduction to each chapter is brief and attempts only to supply general knowledge in the particular field.

Author

Chapter 1

Introduction

NATURE OF ORGANIC CHEMISTRY

Organic chemistry is a specific discipline within the subject of chemistry. It is the scientific study of the structure, properties, composition, reactions, and preparation (by synthesis or by other means) of chemical compounds of carbon and hydrogen, which may contain any number of other elements, such as nitrogen, oxygen, halogens, and, more rarely, phosphorus or sulfur.

Organic chemicals get their diversity from the many different ways carbon can bond to other atoms. Carbon (C) appears in the second row of the periodic table and has four bonding electrons in its valence shell. Similar to other non-metals, carbon needs eight electrons to satisfy its valence shell.

Carbon therefore forms four bonds with other atoms (each bond consisting of one of carbon's electrons and one of the bonding atom's electrons). Every valence electron participates in bonding, thus a carbon atom's bonds will be distributed evenly over the atom's surface. These bonds form a tetrahedron (a pyramid with a spike at the top), as illustrated below:

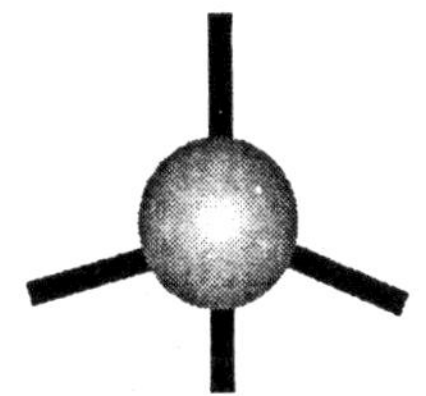

Fig. Carbon Forms 4 Bonds

The simplest organic chemicals, called hydrocarbons, contain only carbon and hydrogen atoms; the simplest hydrocarbon (called methane) contains a single carbon atom bonded to four hydrogen atoms:

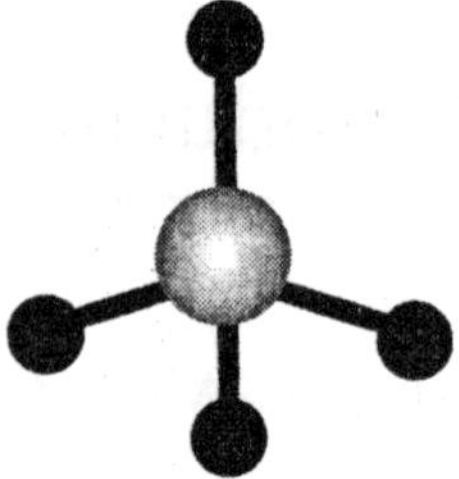

Fig. Methane - a Carbon Atom Bonded to 4 Hydrogen Atoms

But carbon can bond to other carbon atoms in addition to hydrogen, as illustrated in the molecule ethane below:

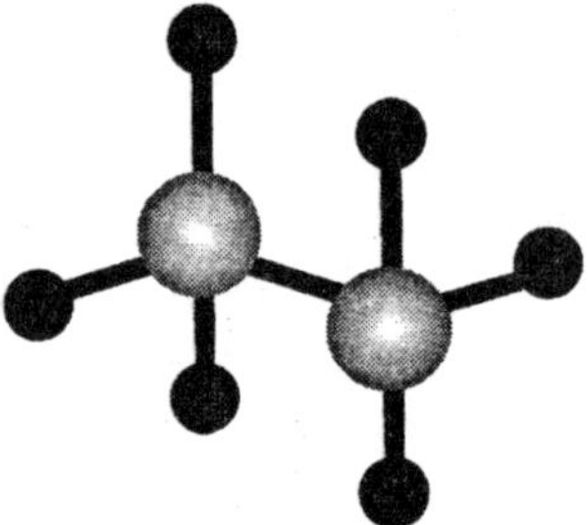

Fig. Ethane - A Carbon-carbon Bond

In fact, the uniqueness of carbon comes from the fact that it can bond to itself in many different ways. Carbon atoms can form long chains:

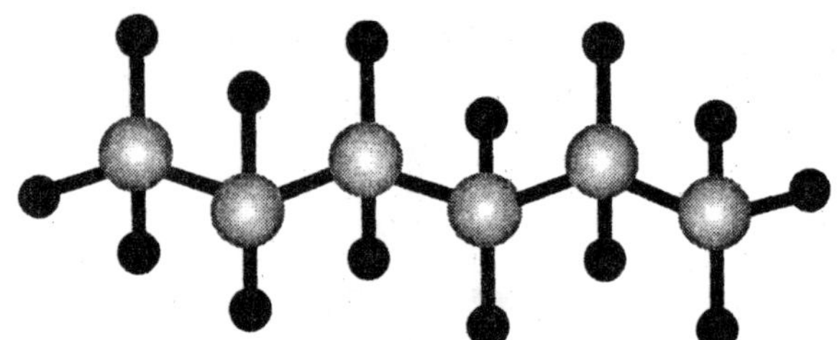

Fig. Hexane - a 6-carbon Chain

Branched Chains:

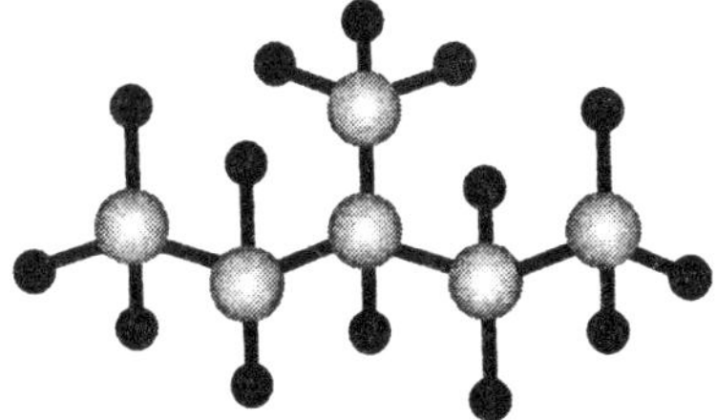

Fig. Isohexane - a Branched-carbon Chain

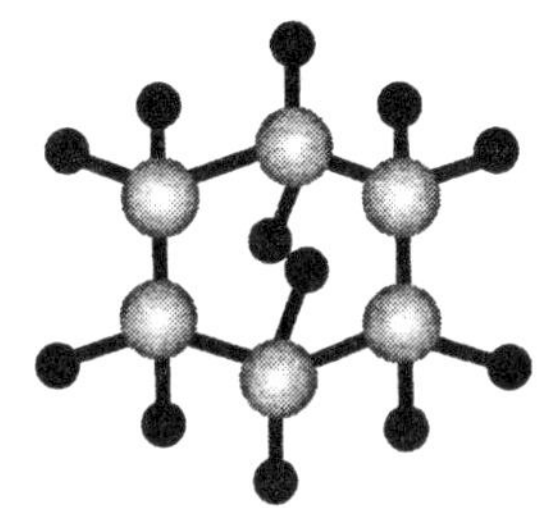

Fig. Cyclohexane - a Ringed Hydrocarbon

There appears to be almost no limit to the number of different structures that carbon can form. To add to the complexity of organic chemistry, neighbouring carbon atoms can form double and triple bonds in addition to single carbon-carbon bonds

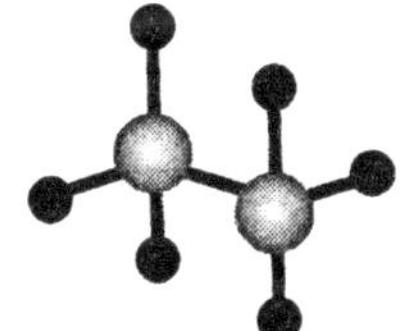

Fig. Sing Bond

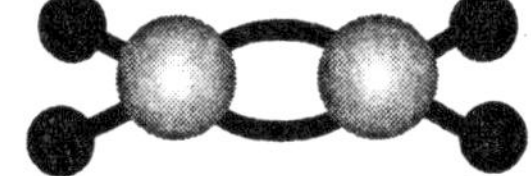

Fig. Double Bond

Fig. Triple Bond

Hydrocarbons are divided into two groups - aliphatic and aromatic. Aliphatic hydrocarbons can be simple like methane (1) (the main constituent of natural gas), which has only one carbon and four hydrogens, or very complex with branched or cyclic structures like cyclohexane.

```
                              H   H
                               \ /
                                C
     H              H\       /     \       /H
     |                C                  C
   H-C-H            H/|                  |\H
     |                |                  |
     H              H\|                  |/H
                      C                  C
    (1)       (2)   H/       \     /       \H
                                C
                               / \
                              H   H
```

Aliphatic compounds can be divided into categories depending on structural features. They can be:

- Alkanes which have only single bonds between carbon atoms

```
  H H
  | |
H-C-C-H
  | |
  H H
```

- Alkanes which have one double bond

```
H-C-C-H
  | |
  H H
```

- Alkynes which have one triple bond

```
H-C-C-H
```

- Cyclic (the carbons are joined to form one or more circles)

```
              H   H
               \ /
                C
    H\       /     \       /H
      C                  C
    H/|                  |\H
      |                  |
    H\|                  |/H
      C                  C
    H/       \     /       \H
                C
               / \
              H   H
```

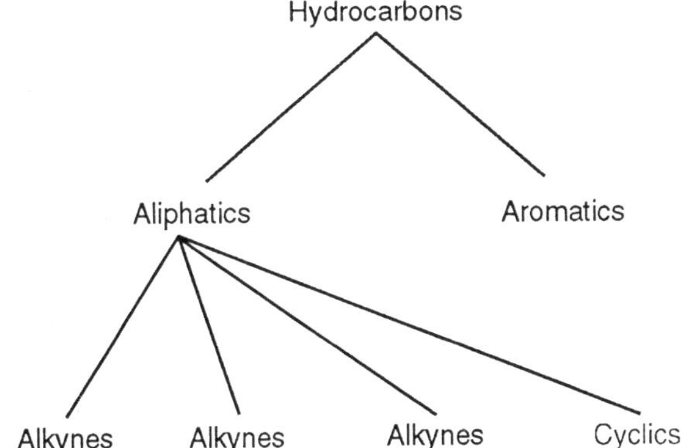

Aliphatic compounds can be represented by drawing the full structures or they can be written in an abbreviated form such as CH_4 for methane (1) and $CH_3CH_2CH_3$ for propane .

FUNCTIONAL GROUPS

Functional groups are specific groups of atoms within molecules, that are responsible for the characteristic chemical reactions of those molecules. The same functional group will undergo the same or similar chemical reaction(s) regardless of the size of the molecule it is a part of. The term 'functional' group is linked to the concept of a homologous series. A homologous series is a group of molecules with the same general formula and the same functional group. They have similar physical and chemical properties (albeit with trends e.g. increasing boiling point with increasing carbon chain length). The terms higher/lower refer to a larger/smaller carbon chain.

Many important organic chemistry molecules contain oxygen or nitrogen. It's a good idea to memorize the names and structures of these functional groups.

–X, X= Cl, Br, I, F	Alkyl Halide Functional Group
–OH	Alcohol Functional Group
–O–	Ether Functional Group
—C—H || C	Aldehyde Functional Group

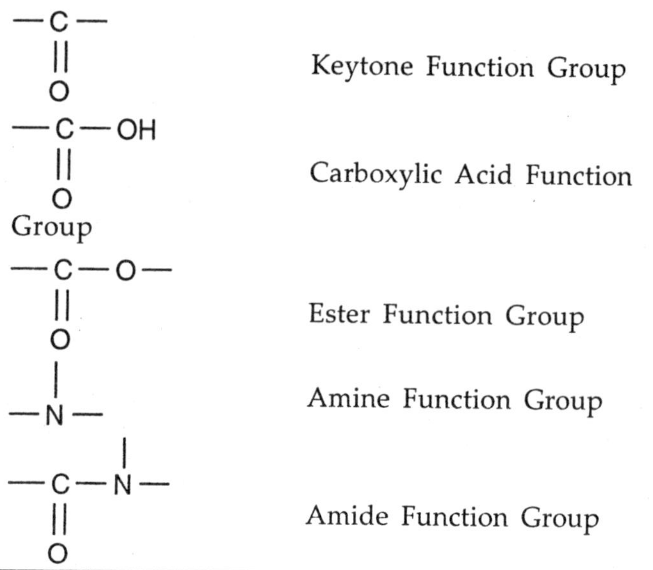

IUPAC NOMENCLATURE

The IUPAC nomenclature of organic chemistry is a systematic way of naming organic chemical compounds as recommended by the International Union of Pure and Applied Chemistry (IUPAC). Ideally, every organic compound should have a name from which an unambiguous structural formula can be drawn.

For ordinary communication, to spare a tedious description, the official IUPAC naming recommendations are not always followed in practice except when it is necessary to give a concise definition to a compound, or when the IUPAC name is simpler (viz. ethanol against ethyl alcohol). Otherwise the common or trivial name may be used, often derived from the source of the compound.

Nomenclature of Saturated Hydrocarbons

- Select the longest continuous chain of carbon atoms in the molecule. The compound is named as a derivative of this alkane
- Number the carbon atoms in the parent chain

starting from the end which gives lowest possible sum for the numbers of the carbon atoms carrying the substituents

- That set of locants is preferred, which when compared term by term with other set of locants, each in order of increasing magnitude, has the lowest term at the first point of difference. For example the set of locants (2,7,8) is preferred over the set of locants (3,4,9) since 2 comes before 3 even though the sum of locants in the former case is 17 while in the latter case, it is 16

$$\begin{array}{ccccccccccccccccccc} 1 & & 2 & & 3 & & 4 & & 5 & & 6 & & 7 & & 8 & & 9 & & 10 \\ Ch_3 & - & Ch & - & Ch_2 & - & Ch_2 & - & Ch_2 & - & Ch_2 & - & Ch & - & Ch & - & Ch_2 & - & Ch_3 \\ & & | & & & & & & & & & & | & & | & & & & \\ & & Ch_3 & & & & & & & & & & Ch_3 & & Ch_3 & & & & \end{array}$$

- The correct name is: 2,7,8 - Trimethyldecane and not 3,4,9 - Trimethyldecane

Nomenclature of compounds containing functional group or multiple bonds:

- Select the longest continuous chain containing the carbon atoms having the functional group or those involved in the multiple bonds
- The numbering of atoms in the parent chain is done in such a way that the carbon atom bearing the functional group or those carrying the multiple bond gets the lowest possible number
- While writing the name of alkene (double bond) or alkyne (triple bond), the primary suffix 'ane' of the corresponding alkane is replaced by 'ene' and 'yne' respectively. However, if the multiple bond occurs twice or thrice in the parent chain, the prefix di- or tri- is attached to the primary suffix ene or yne
- In naming the organic compounds containing one functional group a suffix known as secondary suffix is added to the primary suffix (giving

number of carbon atoms in the chain) to indicate the nature of the functional group. A few important secondary suffixes are:

Functional group	Secondary suffix	Functional group	Secondary suffix
Alcohols (-OH)	-ol	Aldehydes (-CHO)	-al
Ketones (>C=O)	-one	Carboxylic acids (-COOH)	-oic acid
Amines ($-NH_2$)	-amine	Acid amides ($-CONH_2$)	-amide
Acid chlorides (-COCL)	-oyl chloride	Esters (-COOR)	-oate
Nitrites (-C≡N)	-nitrite	Thioalcohols (-SH)	-thiol

Nomenclature of compounds having polyfunctional groups:

When an organic compound contains two or more functional groups, one group is called the principal functional group while the others are called the secondary functional groups and are treated as substituents: The order of preference for principal group is: Carboxylic acid > acid anhydrides > esters > acid halides > amides > nitrites > aldehydes > ketone > alcohols > amines > double bond > triple bond.

When the functional groups act as substituents, they ar named as:

Functional group	Prefix	Functional group	Prefix
- COOH	Carboxy	-CHO	Formyl
-COOR	Alkoxy cabonyl or Carbalkoxy	>CO	Oxo or Keto
-COCL	Chloroformyl	-OH	Hydroxy
$-CONH_2$	Carbamoyl	-SH	Mecaplo
-CN	Cyano	$-NH_2$	

Amino		
-OR	Akoxy	=NH
Imino		
-X	Halo	$-NO_2$
Nitro		

Nomenclature of Simple Aromatic Compounds:

- *Nuclear Substituted*: In these the functional group is directly attached to the benzene ring. Most of these compounds are better known by their common and historical names. In the IUPAC system, they are named as derivatives of benzene.
- *Side Chain Substituted*: In these the functional group is present in the side chain of the benzene ring. Both in the common and IUPAC systems, these are usually named as phenyl derivatives of the corresponding aliphatic compounds.

Chapter 2

Structure and Atomic Orbitals

A SIMPLE VIEW

The electronic structures of hydrogen and carbon can be drawn as:

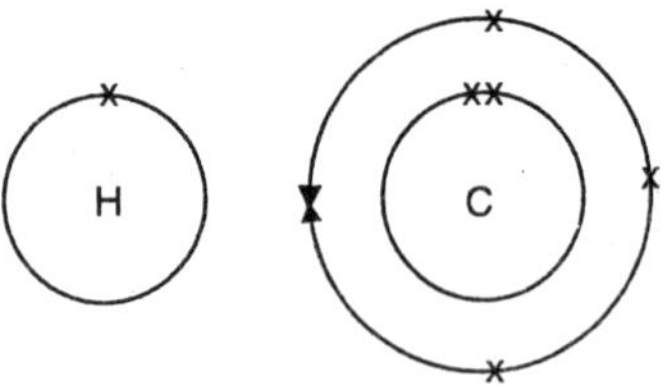

The circles show energy levels - representing increasing distances from the nucleus. You could straighten the circles out and draw the electronic structure as a simple energy diagram.

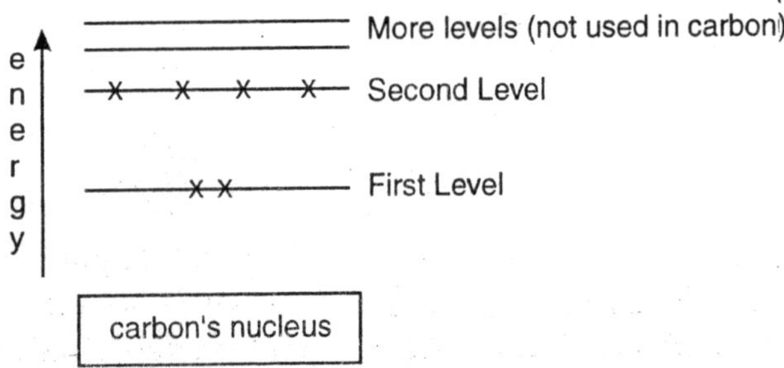

ATOMIC ORBITALS

Orbits and orbitals sound similar, but they have quite different meanings. It is essential that you understand the difference between them.

ORBITS FOR ELECTRONS

To plot a path for something you need to know exactly where the object is and be able to work out exactly where it's going to be an instant later. You can't do this for electrons. The Heisenberg Uncertainty Principle says that you can't know with certainty both where an electron is and where it's going next. That makes it impossible to plot an orbit for an electron around a nucleus.

HYDROGEN'S ELECTRON - THE 1S ORBITAL

Suppose you had a single hydrogen atom and at a particular instant plotted the position of the one electron. Soon afterwards, you do the same thing, and find that it is in a new position. You have no idea how it got from the first place to the second. You keep on doing this over and over again, and gradually build up a sort of 3D map of the places that the electron is likely to be found. In the hydrogen case, the electron can be found anywhere within a spherical space surrounding the nucleus. The diagram shows a *cross-section* through this spherical space. 95% of the time (or any other per centage you choose), the electron will be found within a fairly easily defined region of space quite close to the nucleus. Such a region of space is called an orbital. You can think of an orbital as being the region of space in which the electron lives. What is the electron doing in the orbital? We don't know, we can't know, and so we just ignore the problem! All you can say is that if an electron is in a particular orbital it will have a particular definable energy.

Each Orbital has a name.

Fig. 1s orbital

The orbital occupied by the hydrogen electron is called a 1s orbital. The "1" represents the fact that the orbital is in the energy level closest to the nucleus. The "s" tells you about the shape of the orbital. s orbitals are spherically symmetric around the nucleus - in each case, like a hollow ball made of rather chunky material with the nucleus at its centre.

Fig. 2d orbital

The orbital above is a 2s orbital. This is similar to a 1s orbital except that the region where there is the greatest chance of finding the electron is further from the nucleus - this is an orbital at the second energy level.

If you look carefully, you will notice that there is another region of slightly higher electron density (where the dots are thicker) nearer the nucleus. ("Electron density" is another way of talking about how likely you are to find an electron at a particular place.) 2s (and 3s, 4s, etc) electrons spend some of their time closer to the nucleus than you might expect. The effect of this is to slightly reduce the energy of electrons in s orbitals. The nearer the nucleus the electrons get, the lower their energy. 3s, 4s (etc) orbitals get progressively further from the nucleus.

ORBITALS

Not all electrons inhabit s orbitals (in fact, very few electrons live in s orbitals). At the first energy level, the only orbital available to electrons is the 1s orbital, but at the second level, as well as a 2s orbital, there are also orbitals called 2p orbitals.

Fig. p Orbital

A p orbital is rather like 2 identical balloons tied together at the nucleus. The diagram on the right is a cross-section through that 3-dimensional region of space. Once again, the orbital shows where there is a 95% chance of finding a particular electron.

Unlike an s orbital, a p orbital points in a particular direction. At any one energy level it is possible to have three absolutely equivalent p orbitals pointing mutually at right angles to each other. These are arbitrarily given the symbols p_x, p_y and p_z. This is simply for convenience - what you might think of as the x, y or z direction changes constantly as the atom tumbles in space. The p orbitals at the second energy level are called $2p_x$, $2p_y$ and $2p_z$. There are similar orbitals at subsequent levels - $3p_x$, $3p_y$, $3p_z$, $4p_x$, $4p_y$, $4p_z$ and so on.

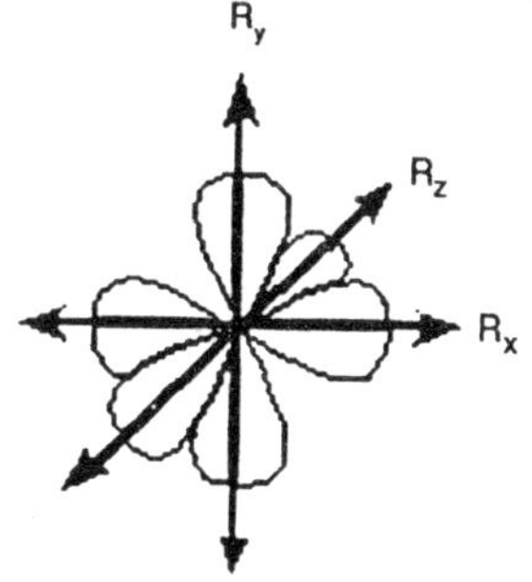

All levels except for the first level have p orbitals. At the higher levels the lobes get more elongated, with the most likely place to find the electron more distant from the nucleus.

Fitting electrons into orbitals

Because for the moment we are only interested in the electronic structures of hydrogen and carbon, we don't need to concern ourselves with what happens beyond the second energy level.

Remember

At the first level there is only one orbital - the 1s orbital.

At the second level there are four orbitals - the 2s, $2p_x$, $2p_y$ and $2p_z$ orbitals. Each orbital can hold either 1 or 2 electrons, but no more.

"Electrons-in-boxes"

Orbitals can be represented as boxes with the electrons in them shown as arrows. Often an up-arrow and a down-arrow are used to show that the electrons are in some way different.

1s

A 1s orbital holding 2 electrons would be drawn as shown on the right, but it can be written even more quickly as $1s^2$. This is read as "one s two" - not as "one s squared".

You mustn't confuse the two numbers in this notation:

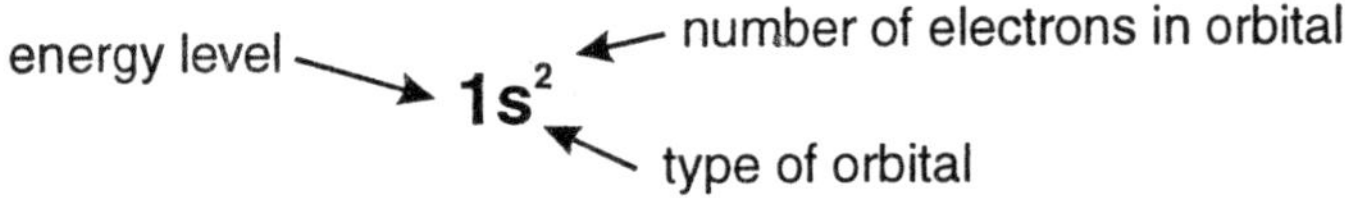

Order of Filling Orbitals

Electrons fill low energy orbitals (closer to the nucleus) before they fill higher energy ones. Where there is a choice between orbitals of equal energy, they fill the orbitals singly as far as possible.

The diagram (not to scale) summarises the energies of the various orbitals in the first and second levels.

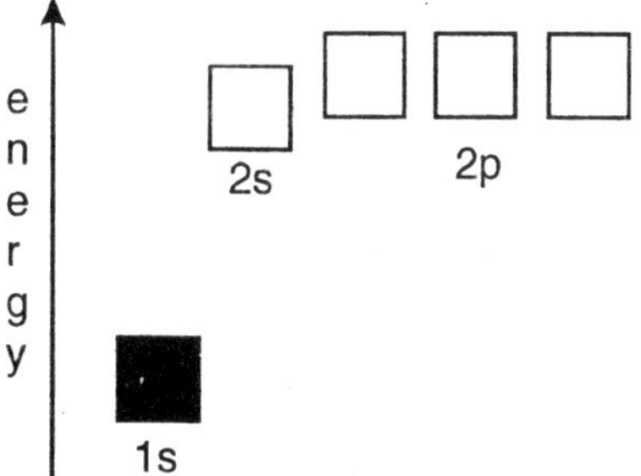

Notice that the 2s orbital has a slightly lower energy than the 2p orbitals. That means that the 2s orbital will fill with electrons before the 2p orbitals. All the 2p orbitals have exactly the same energy.

Electronic Structure of Hydrogen

Hydrogen only has one electron and that will go into the orbital with the lowest energy - the 1s orbital. Hydrogen has an electronic structure of $1s^1$. We have already described this orbital earlier.

Electronic Structure of Carbon

Carbon has six electrons. Two of them will be found in the 1s orbital close to the nucleus. The next two will go into the 2s orbital. The remaining ones will be in two separate 2p orbitals. This is because the p orbitals all have the same energy and the electrons prefer to be on their own if that's the case.

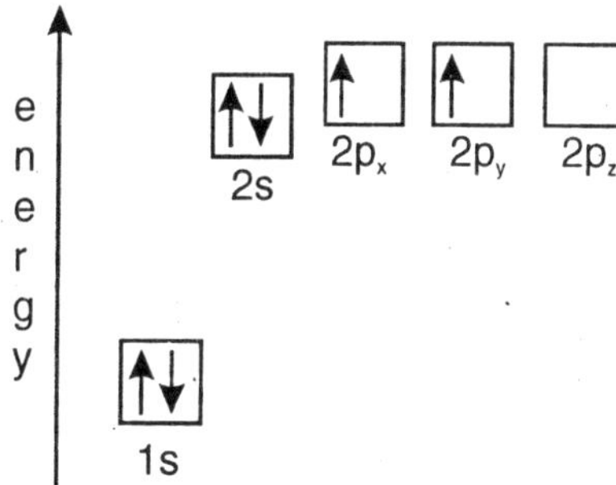

The electronic structure of carbon is normally written $1s^2 2s^2 2p_x{}^1 2p_y{}^1$.

Chapter 3

Organic Molecules

ELECTRONEGATIVITY

DEFINITION

Electronegativity is a measure of the tendency of an atom to attract a bonding pair of electrons. The Pauling scale is the most commonly used. Fluorine (the most electronegative element) is given a value of 4.0, and values range down to caesium and francium which are the least electronegative at 0.7.

RESULT OF ELECTRONEGATIVITY

The most obvious example of this is the bond between two carbon atoms. Both atoms will attract the bonding pair to exactly the same extent. That means that on average the electron pair will be found half way between the two nuclei, and you could draw a picture of the bond like this:

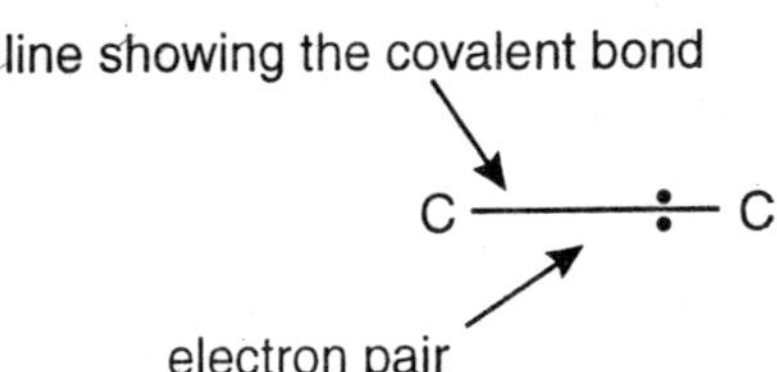

It is important to realise that this is an average picture. The electrons are actually in a sigma orbital, and are moving constantly within that orbital.

Carbon-fluorine Bond

Fluorine is much more electronegative than carbon. The actual values on the Pauling scale are,

Carbon 2.5 Fluorine 4.0

That means that fluorine attracts the bonding pair much more strongly than carbon does. The bond - on average - will look like this:

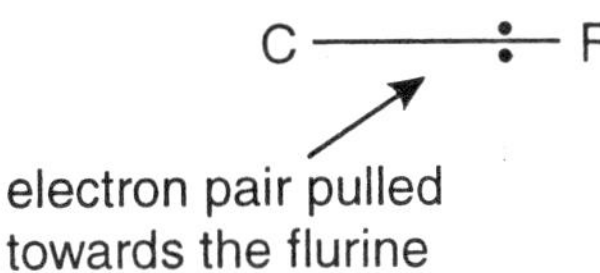

Why is fluorine more electronegative than carbon?

A simple dots-and-crosses diagram of a C-F bond is perfectly adequate to explain it.

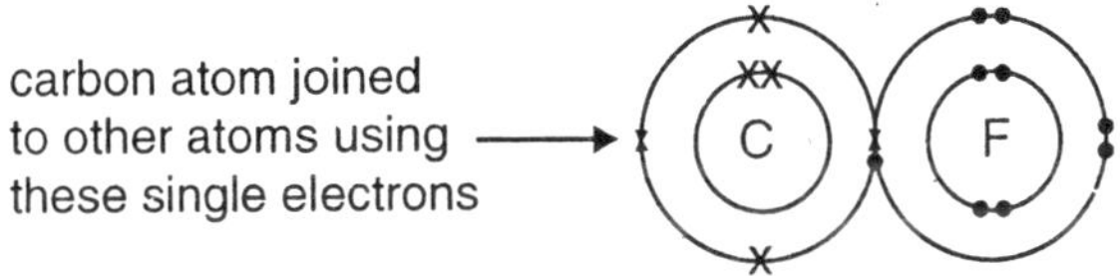

The bonding pair is in the second energy level of both carbon and fluorine, so in the absence of any other effect, the distance of the pair from both nuclei would be the same. The electron pair is shielded from the full force of both nuclei by the 1s electrons - again there is nothing to pull it closer to one atom than the other.

BUT, the fluorine nucleus has 9 protons whereas the carbon nucleus has only 6.

Allowing for the shielding effect of the 1s electrons, the bonding pair feels a net pull of about 4+ from the carbon, but about 7+ from the fluorine. It is this extra nuclear charge which pulls the bonding pair (on average) closer to the fluorine than the carbon.

Carbon-chlorine Bond

The electronegativities are:

Carbon 2.5

Chlorine 3.0

The bonding pair of electrons will be dragged towards the chlorine but not as much as in the fluorine case. Chlorine isn't as electronegative as fluorine.

Why isn't Chlorine as Electronegative as Fluorine?

Chlorine is a bigger atom than fluorine.

Fluorine: $1s^2 2s^2 2p_x^2 2p_y^2 2p_z^1$

Chlorine: $1s^2 2s^2 2p_x^2 2p_y^2 2p_z^2 3s^2 3p_x^2 3p_y^2 3p_z^1$

In the chlorine case, the bonding pair will be shielded by all the 1-level and 2-level electrons. The 17 protons on the nucleus will be shielded by a total of 10 electrons, giving a net pull from the chlorine of about 7+.

That is the same as the pull from the fluorine, but with chlorine the bonding pair starts off further away from the nucleus because it is in the 3-level. Since it is further away, it feels the pull from the nucleus less strongly.

POLAR BONDS

Think about the carbon-fluorine bond again. Because the bonding pair is pulled towards the fluorine end of the bond, that end is left rather more negative than it would otherwise be. The carbon end is left rather short of electrons and so becomes slightly positive.

$$\overset{\delta+}{\mathrm{C}} \text{———}{:}\overset{\delta-}{\mathrm{F}}$$

The symbols δ+ and - mean "slightly positive" and "slightly negative". You read + as "delta plus" or "delta positive". We describe a bond having one end slightly positive and the other end slightly negative as being *polar*.

Inductive Effects

An atom like fluorine which can pull the bonding pair away from the atom it is attached to is said to have a negative inductive effect. Most atoms that you will come across have a negative inductive effect when they are attached to a carbon atom, because they are mostly more electronegative than carbon. You will come across some groups of atoms which have a slight positive inductive effect - they "push" electrons towards the carbon they are attached to, making it slightly negative.

Inductive effects are sometimes given symbols: *-I* (a negative inductive effect) and *+I* (a positive inductive effect).

Examples of Polar Bonds

Hydrogen bromide (and other hydrogen halides),

$$\overset{\delta+}{\mathrm{H}}\text{———}\overset{\delta-}{\mathrm{Br}}$$

Bromine (and the other halogens) are all more electronegative than hydrogen, and so all the hydrogen halides have polar bonds with the hydrogen end slightly positive and the halogen end slightly negative. The polarity of these molecules is important in their reactions with alkenes.

Carbon-Bromine Bond in Halogenoalkanes

Bromine is more electronegative than carbon and so the bond is polarized in the way that we have already described with C-F and C-Cl.

$$\overset{\delta+}{\mathrm{C}}\text{———}\overset{\delta-}{\mathrm{Br}}$$

The polarity of the carbon-halogen bonds is important in the reactions of the halogenoalkanes.

Carbon-oxygen Double Bond

An orbital model of the C=O bond in methanal, HCHO, looks like this:

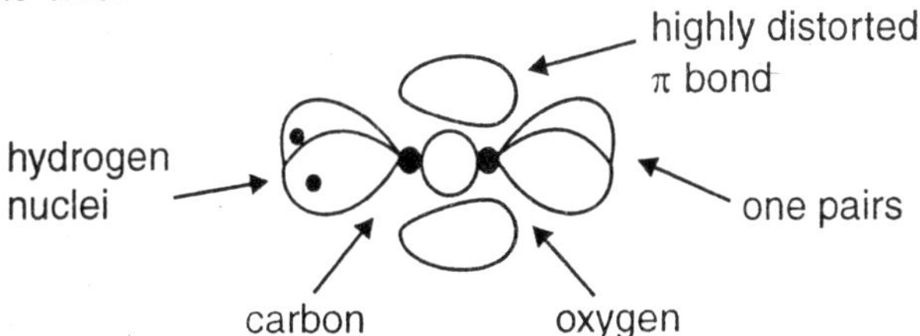

The very electronegative oxygen atom pulls both bonding pairs towards itself - in the sigma bond and the pi bond. That leaves the oxygen fairly negative and the carbon fairly positive.

$$\overset{\delta+}{\mathrm{C}}\text{═══}\overset{\delta-}{\mathrm{O}}$$

Chapter 4

Proteins

OVERVIEW

Proteins are the most versatile macromolecules in living systems and serve crucial functions in essentially all biological processes. They function as catalysts, they transport and store other molecules such as oxygen, they provide mechanical support and immune protection, they generate movement, they transmit nerve impulses, and they control growth and differentiation. Indeed, much of this text will focus on understanding what proteins do and how they perform these functions.

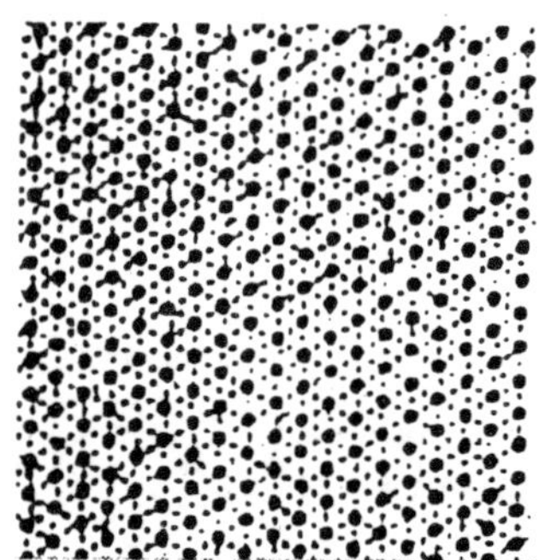

Fig. A Complex Protein Assembly

Several key properties enable proteins to participate in such a wide range of functions:

- *Proteins are linear polymers built of monomer units called amino acids.* The construction of a vast array of macromolecules from a limited number of monomer building blocks is a recurring theme in biochemistry. Does protein function depend on the linear sequence of amino acids? The function of a protein is directly dependent on its threedimensional structure. Remarkably, proteins spontaneously fold up into three-dimensional structures that are determined by the sequence of amino acids in the protein polymer. Thus, *proteins are the embodiment of the transition from the one-dimensional world of sequences to the three-dimensional world of molecules capable of diverse activities.*

- *Proteins contain a wide range of functional groups.* These functional groups include alcohols, thiols, thioethers, carboxylic acids, carboxamides, and a variety of basic groups. When combined in various sequences, this array of functional groups accounts for the broad spectrum of protein function. For instance, the chemical reactivity associated with these groups is essential to the function of *enzymes,* the proteins that catalyze specific chemical reactions in biological systems.
- *Proteins can interact with one another and with other biological macromolecules to form complex assemblies.* The proteins within these assemblies can act synergistically to generate capabilities not afforded by the individual component proteins. These assemblies include macro-molecular machines that carry out the accurate replication of DNA, the transmission of signals within cells, and many other essential processes.
- *Some proteins are quite rigid, whereas others display limited flexibility.* Rigid units can function as structural elements in the cytoskeleton (the internal scaffolding within cells) or in connective tissue. Parts of proteins with limited flexibility may act as hinges, springs, and levers that are crucial to protein function, to the assembly of proteins with one another and with other molecules into complex units, and to the transmission of information within and between cells

ORIGINATION

Amino acids are the building blocks of proteins. An *α-amino acid* consists of a central carbon atom, called the *α carbon,* linked to an amino group, a carboxylic acid group, a hydrogen atom, and a distinctive R group.

The R group is often referred to as the *side chain*. With four different groups connected to the tetrahedral α-carbon atom, α-amino acids are *chiral;* the two mirror-image forms are called the l isomer and the d isomer.

Only l amino acids are constituents of proteins. For almost all amino acids, the l isomer has *S* (rather than *R*) absolute configuration.

Although considerable effort has gone into understanding why amino acids in proteins have this absolute configuration, no satisfactory explanation has been arrived at. It seems plausible that the selection of l over d was arbitrary but, once made, was fixed early in evolutionary history.

Amino acids in solution at neutral pH exist predominantly as *dipolar ions* (also called *zwitterions*). In the dipolar form, the amino group is protonated ($-NH_3^+$) and the carboxyl group is deprotonated ($-COO^-$).

The ionization state of an amino acid varies with pH. In acid solution (e.g., pH 1), the amino group is protonated ($-NH_3^+$) and the carboxyl group is not dissociated (-COOH). As the pH is raised, the carboxylic acid is the first group to give up a proton, inasmuch as its pK_a is near 2. The dipolar form persists until the pH approaches 9, when the protonated amino group loses a proton.

Twenty kinds of side chains varying in *size, shape, charge, hydrogen-bonding capacity, hydrophobic character,* and *chemical reactivity* are commonly found in proteins. Indeed, all proteins in all species—bacterial, archaeal, and eukaryotic—are constructed from the same set of 20 amino acids.

This fundamental alphabet of proteins is several billion years old. The remarkable range of functions mediated by proteins results from the diversity and versatility of these 20 building blocks.

Let us look at this set of amino acids. The simplest one is *glycine,* which has just a hydrogen atom as its side chain. With two hydrogen atoms bonded to the α-carbon atom, glycine is unique in being *achiral. Alanine,* the next simplest amino acid, has a methyl group (-CH_3) as its side chain.

Larger hydrocarbon side chains are found in *valine, leucine,* and *isoleucine. Methionine* contains a largely *aliphatic* side chain that includes a *thioether* (-S-) group. The side chain of isoleucine includes an additional chiral centre; only the isomer is found in proteins. The larger aliphatic side chains are *hydrophobic*—that is, they tend to cluster together rather than contact water.

The three-dimensional structures of water-soluble proteins are stabilized by this tendency of hydrophobic groups to come together, called *the hydrophobic effect.* The different sizes and shapes of these hydrocarbon side chains enable them to pack together to form compact structures with few holes.

Proline also has an aliphatic side chain, but it differs from other members of the set of 20 in that its side chain is bonded to both the nitrogen and the α-carbon atoms. Proline markedly influences protein architecture because its ring structure makes it more conformationally restricted than the other amino acids.

Three amino acids with relatively simple *aromatic side chains* are part of the fundamental repertoire. *Phenylalanine,* as its name indicates, contains a phenyl ring attached in place of one of the hydrogens of alanine. The aromatic ring of *tyrosine* contains a hydroxyl group.

This hydroxyl group is reactive, in contrast with the rather inert side chains of the other amino acids discussed thus far. *Tryptophan* has an indole ring joined to a methylene (-CH_2-) group; the indole group comprises two fused rings and an NH group. Phenylalanine is purely hydrophobic, whereas tyrosine and tryptophan are less so because of their hydroxyl and NH groups. The aromatic rings of tryptophan and tyrosine contain delocalized p electrons that strongly absorb ultraviolet light.

A compound's *extinction coefficient* indicates its ability to absorb light. Beer's law gives the absorbance (*A*) of light at a given wavelength:

$A = \varepsilon cl$ Beer's Law

where ε is the extinction coefficient [in units that are the reciprocals of molarity and distance in centimetres (M^{-1} cm^{-1})], *c* is the concentration of the absorbing species (in units of molarity, M), and *l* is the length through which the light passes (in units of centimetres).

For tryptophan, absorption is maximum at 280 nm and the extinction coefficient is 3400 M^{-1} cm^{-1} whereas, for tyrosine, absorption is maximum at 276 nm and the extinction coefficient is a less-intense 1400 M^{-1} cm^{-1}. Phenylalanine absorbs light less strongly and at shorter wavelengths. The absorption of light at 280 nm can be used to estimate the concentration of a protein in solution if the number of tryptophan and tyrosine residues in the protein is known.

Two amino acids, *serine* and *threonine,* contain aliphatic *hydroxyl groups*. Serine can be thought of as a hydroxylated version of alanine, whereas threonine resembles valine with a hydroxyl group in place of one of the valine methyl groups. The hydroxyl groups on serine and threonine make them much more *hydrophilic* (water loving) and *reactive* than alanine and valine. Threonine, like isoleucine, contains an additional asymmetric centre; again only one isomer is present in proteins.

Cysteine is structurally similar to serine but contains a *sulfhydryl,* or *thiol* (-SH), group in place of the hydroxyl (-OH) group. The sulfhydryl group is much more reactive. Pairs of sulfhydryl groups may come together to form disulfide bonds, which are particularly important in stabilizing some proteins, as will be discussed shortly.

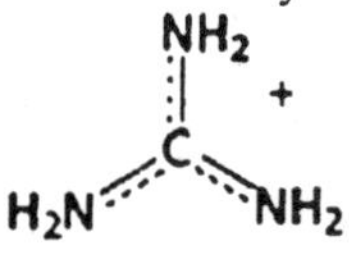

Guanidinium 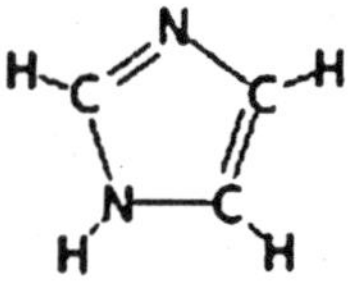 Imidazole

We turn now to amino acids with very polar side chains that render them highly hydrophilic. *Lysine* and *arginine* have relatively long side chains that terminate with groups that are *positively charged* at neutral pH. Lysine is capped by a primary amino group and arginine by a guanidinium group. *Histidine* contains an imidazole group, an aromatic ring that also can be positively charged.

With a pK_a value near 6, the imidazole group can be uncharged or positively charged near neutral pH, depending on its local environment. Indeed, histidine is often found in the active sites of enzymes, where the imidazole ring can bind and release protons in the course of enzymatic reactions.

The set of amino acids also contains two with *acidic side chains: aspartic acid* and *glutamic acid.* These amino acids are often called *aspartate* and *glutamate* to emphasize that their side chains are usually negatively charged at physiological pH. Nonetheless, in some proteins these side chains do accept protons, and this ability is often functionally important. In addition, the set includes uncharged derivatives of aspartate and glutamate—*asparagine* and *glutamine*—each of which contains a terminal *carboxamide* in place of a carboxylic acid.

Seven of the 20 amino acids have readily ionizable side chains. These 7 amino acids are able to donate or accept protons to facilitate reactions as well as to form ionic bonds.

Amino acids are often designated by either a three-letter abbreviation or a one-letter symbol. The abbreviations for amino acids are the first three letters of their names, except for asparagine (Asn), glutamine (Gln), isoleucine (Ile), and tryptophan (Trp). The symbols for many amino acids are the first letters of their names (e.g., G for glycine and L for leucine); the other symbols have been agreed on by convention. These abbreviations and symbols are an integral part of the vocabulary of biochemists.

How did this particular set of amino acids become the building blocks of proteins? First, as a set, they are diverse; their structural and chemical properties span a wide range, endowing proteins with the versatility to assume many functional roles. Second, as noted many of these amino acids were probably available from prebiotic reactions. Finally, excessive intrinsic reactivity may have eliminated other possible amino acids.

For example, amino acids such as homoserine and homocysteine tend to form five-membered cyclic forms that limit their use in proteins; the alternative amino acids that are found in proteins—serine and cysteine—do not readily cyclize, because the rings in their cyclic forms are too small.

Proteins are the workhorses of biochemistry, participating in essentially all cellular processes. Protein structure can be described at four levels. The primary structure refers to the amino acid sequence. The secondary structure refers to the conformation adopted by local regions of the polypeptide chain.

Tertiary structure describes the overall folding of the polypeptide chain. Finally, quaternary structure refers to the specific association of multiple polypeptide chains to form multisubunit complexes.

PROTEINS AS LINEAR POLYMERS

Proteins are linear polymers of amino acids. Each amino acid consists of a central tetrahedral carbon atom linked to an amino group, a carboxylic acid group, a distinctive side chain, and a hydrogen. These tetrahedral centres, with the exception of that of glycine, are chiral; only the l isomer exists in natural proteins. All natural proteins are constructed from the same set of 20 amino acids. The side chains of these 20 building blocks vary tremendously in size, shape, and the presence of functional groups.

They can be grouped as follows:

- *Aliphatic side chains*: glycine, alanine, valine, leucine, isoleucine, methionine, and proline;
- *Aromatic side chains*: phenylalanine, tyrosine, and tryptophan;
- *Hydroxyl-containing aliphatic side chains*: serine and threonine;
- *Sulfhydryl*: containing cysteine;

- *Basic side chains*: lysine, arginine, and histidine;
- *Acidic side chains*: aspartic acid and glutamic acid; and
- *Carboxamide-containing side chains*: asparagine and glutamine. These groupings are somewhat arbitrary and many other sensible groupings are possible.

PRIMARY STRUCTURE

The amino acids in a polypeptide are linked by amide bonds formed between the carboxyl group of one amino acid and the amino group of the next.

This linkage, called a peptide bond, has several important properties. First, it is resistant to hydrolysis so that proteins are remarkably stable kinetically. Second, the peptide group is planar because the C-N bond has considerable double-bond character. Third, each peptide bond has both a hydrogen-bond donor (the NH group) and a hydrogen-bond acceptor (the CO group).

Hydrogen bonding between these backbone groups is a distinctive feature of protein structure. Finally, the peptide bond is uncharged, which allows protcins to form tightly packed globular structures having significant amounts of the backbone buried within the protein interior. Because they are linear polymers, proteins can be described as sequences of amino acids. Such sequences are written from the amino to the carboxyl terminus.

SECONDARY STRUCTURE

Polypeptide Chains Can Fold into Regular Structures Such as the Alpha Helix, the Beta Sheet, and Turns and Loops Two major elements of secondary structure are the α helix and the β strand. In the *β* helix, the polypeptide chain twists into a tightly packed rod.

Within the helix, the CO group of each amino acid is hydrogen bonded to the NH group of the amino acid four residues along the polypeptide chain. In the β strand, the polypeptide chain is nearly fully extended. Two or more β strands connected by NH-to-CO hydrogen bonds come together to form β sheets.

TERTIARY STRUCTURE

Water-Soluble Proteins Fold into Compact Structures with Nonpolar Cores. The compact, asymmetric structure that individual polypeptides attain is called tertiary structure.

The tertiary structures of water-soluble proteins have features in common:

- An interior formed of amino acids with hydrophobic side chains and
- A surface formed largely of hydrophilic amino acids that interact with the aqueous environment. The driving force for the formation of the tertiary structure of water-soluble proteins is the hydrophobic interactions between the interior residues. Some proteins that exist in a hydrophobic environment, in membranes, display the inverse distribution of hydrophobic and hydrophilic amino acids. In these proteins, the hydrophobic amino acids are on the surface to interact with the environment, whereas the hydrophilic groups are shielded from the environment in the interior of the protein.

QUATERNARY STRUCTURE

Polypeptide Chains Can Assemble into Multisubunit Structures. Proteins consisting of more than one polypeptide chain display quaternary structure, and each individual polypeptide chain is called a subunit. Quaternary structure can be as simple as two identical subunits or as complex as dozens of different subunits. In most cases, the subunits are held together by noncovalent bonds.

STRUCTURE OF AMINO ACID

The amino acid sequence completely determines the three-dimensional structure and, hence, all other properties of a protein. Some proteins can be unfolded completely yet refold efficiently when placed under conditions in which the folded form of the protein is stable.

The amino acid sequence of a protein is determined by the sequences of bases in a DNA molecule. This one-dimensional sequence information is extended into the three-dimensional world by the ability of proteins to fold spontaneously. Protein folding is a highly cooperative process; structural intermediates between the unfolded and folded forms do not accumulate.

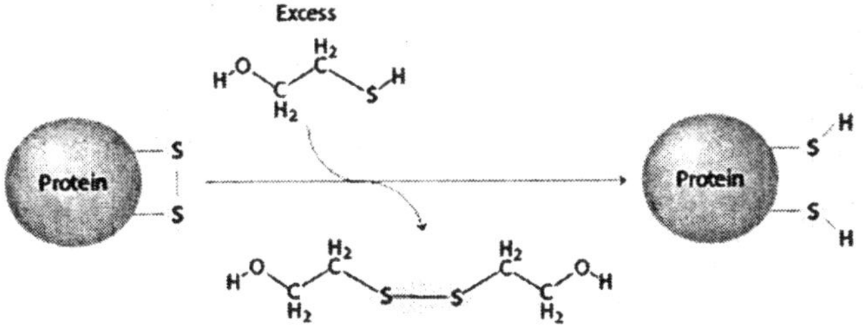

Fig. Role of β -Mercaptoethanol in Reducing Disulfide Bonds

The versatility of proteins is further enhanced by covalent modifications. Such modifications can incorporate functional groups not present in the 20 amino acids. Other modifications are important to the regulation of protein activity. Through their structural stability, diversity, and chemical reactivity, proteins make possible most of the key processes associated with life.

PRIMARY STRUCTURE

Amino Acids Are Linked by Peptide Bonds to Form Polypeptide Chains. Proteins are *linear polymers* formed by linking the α-carboxyl group of one amino acid to the α-amino group of another amino acid with a *peptide bond* (also called an *amide bond*). The formation of a dipeptide from two amino acids is accompanied by the loss of a water molecule. The equilibrium of this reaction lies on the side of hydrolysis rather than synthesis. Hence, the biosynthesis of peptide bonds requires an input of free energy. Nonetheless, peptide bonds are quite *stable kinetically;* the lifetime of a peptide bond in aqueous solution in the absence of a catalyst approaches 1000 years.

A series of amino acids joined by peptide bonds form a *polypeptide chain,* and each amino acid unit in a polypeptide is called a *residue. A polypeptide chain has polarity* because its ends are different, with an α-amino group at one end and an α-carboxyl group at the other. By convention, *the amino end is taken to be the beginning of a polypeptide chain,* and so the sequence of amino acids in a polypeptide chain is written starting with the aminoterminal residue. Thus, in the pentapeptide Tyr-Gly-Gly-Phe-Leu (YGGFL), phenylalanine is the amino-terminal (N-terminal) residue and leucine is the carboxyl-terminal (C-terminal) residue. Leu-Phe-Gly-Gly-Tyr (LFGGY) is a different pentapeptide, with different chemical properties.

A polypeptide chain consists of a regularly repeating part, called the *main chain* or *backbone,* and a variable part, comprising the distinctive *side chains.* The polypeptide backbone is rich in hydrogen-bonding potential. Each residue contains a carbonyl group, which is a good hydrogen-bond acceptor and, with the exception of proline, an NH group, which is a good hydrogen-bond donor. These groups interact with each other and with functional groups from side chains to stabilize particular structures, as will be discussed in detail.

Most natural polypeptide chains contain between 50 and 2000 amino acid residues and are commonly referred to as *proteins.* Peptides made of small numbers of amino acids are called *oligopeptides* or simply *peptides.* The mean molecular weight of an amino acid residue is about 110, and so the molecular weights of most proteins are between 5500 and 220,000. We can also refer to the mass of a protein, which is expressed in units of daltons; one *dalton* is equal to one atomic mass unit. A protein with a molecular weight of 50,000 has a mass of 50,000 daltons, or 50 kd (kilodaltons).

In some proteins, the linear polypeptide chain is cross-linked. The most common cross-links are *disulfide bonds,* formed by the oxidation of a pair of cysteine residues. The resulting unit of linked cysteines is called *cystine.* Extracellular proteins often have several disulfide bonds, whereas intracellular proteins usually lack them. Rarely, nondisulfide cross-links derived from other side chains are present in some proteins. For example, collagen fibers in connective tissue are strengthened in this way, as are fibrin blood clots.

PROTEINS HAVE UNIQUE AMINO ACID SEQUENCES THAT ARE SPECIFIED BY GENES

In 1953, Frederick Sanger determined the amino acid sequence of insulin, a protein hormone. *This work is a landmark in biochemistry because it showed for the first time that a protein has a precisely defined amino acid sequence.* Moreover, it demonstrated that insulin consists only of l amino acids linked by peptide bonds between α-amino and α-carboxyl groups. This accomplishment stimulated other scientists to carry out sequence studies of a wide variety of proteins. Indeed, the complete amino acid sequences of more than 100,000 proteins are now known. *The striking fact is that each protein has a unique, precisely defined amino acid sequence.* The amino acid sequence of a protein is often referred to as its *primary structure.*

A series of incisive studies in the late 1950s and early 1960s revealed that the amino acid sequences of proteins are genetically determined. The sequence of nucleotides in DNA, the molecule of heredity, specifies a complementary sequence of nucleotides in RNA, which in turn specifies the amino acid sequence of a protein. In particular, each of the 20 amino acids of the repertoire is encoded by one or more specific sequences of three nucleotides.

Knowing amino acid sequences is important for several reasons. First, knowledge of the sequence of a protein is usually essential to elucidating its mechanism of action (e.g., the catalytic mechanism of an enzyme). Moreover, proteins with novel properties can be generated by varying the sequence of known proteins. Second, amino acid sequences determine the three-dimensional structures of proteins. Amino acid sequence is the link between the genetic message in DNA and the three-dimensional structure that performs a protein's biological function. Analyses of relations between amino acid sequences and three-dimensional structures of proteins are uncovering the rules that govern the folding of polypeptide chains. Third, sequence determination is a component of molecular pathology, a rapidly growing area of medicine. Alterations in amino acid sequence can produce abnormal function and disease. Severe and sometimes fatal diseases, such as sickle-cell anemia and cystic fibrosis, can result from a change in a single amino acid within a protein. Fourth, the sequence of a protein reveals much about its evolutionary history. Proteins resemble one another in amino acid sequence only if they have a common ancestor. Consequently, molecular events in evolution can be traced from amino acid sequences; molecular paleontology is a flourishing area of research.

POLYPEPTIDE CHAINS

Examination of the geometry of the protein backbone reveals several important features. First, *the peptide bond is essentially planar*. Thus, for a pair of amino acids linked by a peptide bond, six atoms lie in the same plane: the α-carbon atom and CO group from the first amino acid and the NH group and α-carbon atom from the second amino acid. The nature of the chemical bonding within a peptide explains this geometric preference. The peptide bond has considerable *double-bond character*, which prevents rotation about this bond.

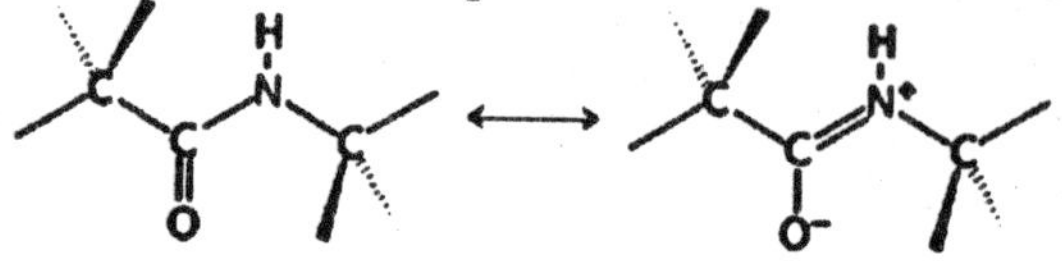

Peptide bond resonance strucutres

The inability of the bond to rotate constrains the conformation of the peptide backbone and accounts for the bond's planarity. This double-bond character is also expressed in the length of the bond between the CO and NH groups. The C-N distance in a peptide bond is typically 1.32 Å, which is between the values expected for a C-N single bond (1.49 Å) and a C=N double bond (1.27 Å). Finally, the peptide bond is uncharged, allowing polymers of amino acids linked by peptide bonds to form tightly packed globular structures.

Two configurations are possible for a planar peptide bond. In the trans configuration, the two α-carbon atoms are on opposite sides of the peptide bond.

In the cis configuration, these groups are on the same

side of the peptide bond. *Almost all peptide bonds in proteins are trans.* This preference for trans over cis can be explained by the fact that steric clashes between groups attached to the α-carbon atoms hinder formation of the cis form but do not occur in the trans configuration. By far the most common cis peptide bonds are X-Pro linkages. Such bonds show less preference for the trans configuration because the nitrogen of proline is bonded to two tetrahedral carbon atoms, limiting the steric differences between the trans and cis forms.

In contrast with the peptide bond, the bonds between the amino group and the α-carbon atom and between the α-carbon atom and the carbonyl group are pure single bonds. The two adjacent rigid peptide units may rotate about these bonds, taking on various orientations. *This freedom of rotation about two bonds of each amino acid allows proteins to fold in many different ways.* The rotations about these bonds can be specified by dihedral angles. The angle of rotation about the bond between the nitrogen and the α-carbon atoms is called *phi* (φ). The angle of rotation about the bond between the α-carbon and the carbonyl carbon atoms is called *psi* (ψ). A clockwise rotation about either bond as viewed from the front of the back group corresponds to a positive value. The φ and y angles determine the path of the polypeptidc chain.

Are all combinations of φ and ψ possible? G. N. Ramachandran recognized that many combinations are forbidden because of steric collisions between atoms. The allowed values can be visualized on a two-dimensional plot called a *Ramachandran diagram.* Three-quarters of the possible (φ, ψ) combinations are excluded simply by local steric clashes. *Steric exclusion, the fact that two atoms cannot be in the same place at the same time, can be a powerful organizing principle.*

The ability of biological polymers such as proteins to fold into welldefined structures is remarkable thermodynamically. Consider the equilibrium between an unfolded polymer that exists as a random coil—that is, as a mixture of many possible conformations—and the folded form that adopts a unique conformation. The favourable entropy associated with the large number of conformations in the unfolded form opposes folding and must be overcome by interactions favouring the folded form. Thus, highly flexible polymers with a large number of possible conformations do not fold into unique structures. The rigidity of the peptide unit and the restricted set of allowed φ and ψ angles limits the number of structures accessible to the unfolded form sufficiently to allow protein folding to occur.

SECONDARY STRUCTURE

Polypeptide Chains Can Fold Into Regular Structures Such as the Alpha Helix, the Beta Sheet, and Turns and Loops. Can a polypeptide chain fold into a regularly repeating structure? In 1951, Linus Pauling and Robert Corey proposed two periodic structures called the *α helix* (*a*lpha helix) and the *β pleated sheet* (*beta* pleated sheet). Subsequently, other structures such as the *β turn* and *omega* (Ω) *loop* were identified. Although not periodic, these common turn or loop structures are well defined and contribute with α helices and β sheets to form the final protein structure.

ALPHA HELIX

In evaluating potential structures, Pauling and Corey considered which conformations of peptides were sterically allowed and which most fully exploited the hydrogen-bonding capacity of the backbone NH and CO groups. The first of their proposed structures, the *α helix*, is a rodlike structure. A tightly coiled backbone forms the inner part of the rod and the side chains extend outward in a helical array. The *α* helix is stabilized by hydrogen bonds between the NH and CO groups of the main chain. In particular, the CO group of each amino acid forms a hydrogen bond with the NH group of the amino acid that is situated four residues ahead in the sequence. Thus, except for amino acids near the ends of an *α* helix, all *the main-chain CO and NH groups are hydrogen bonded*. Each residue is related to the next one by a rise of 1.5 Å along the helix axis and a rotation of 100 degrees, which gives 3.6 amino acid residues per turn of helix. Thus, amino acids spaced three and four apart in the sequence are spatially quite close to one another in an α helix.

In contrast, amino acids two apart in the sequence are situated on opposite sides of the helix and so are unlikely to make contact. The *pitch* of the α helix, which is equal to the product of the translation (1.5 Å) and the number of residues per turn (3.6), is 5.4 Å.

The *screw sense* of a helix can be right-handed (clockwise) or left-handed (counterclockwise). The Ramachandran diagram reveals that both the right-handed and the left-handed helices are among allowed conformations. However, right-handed helices are energetically more favourable because there is less steric clash between the side chains and the backbone. *Essentially all α helices found in proteins are right-handed.* In schematic diagrams of proteins, α helices are depicted as twisted ribbons or rods.

Pauling and Corey predicted the structure of the α helix 6 years before it was actually seen in the x-ray reconstruction of the structure of myoglobin. *The elucidation of the structure of the α helix is a landmark in biochemistry because it demonstrated that the conformation of a polypeptide chain can be predicted if the properties of its components are rigorously and precisely known.*

The α-helical content of proteins ranges widely, from nearly none to almost 100per cent. For example, about 75per cent of the residues in ferritin, a protein that helps store iron, are in α helices. Single α helices are usually less than 45 Å long.

However, two or more α helices can entwine to form a very stable structure, which can have a length of 1000 Å (100 μm, or 0.1μm) or more. Such *α-helical coiled coils* are found in myosin and tropomyosin in muscle, in fibrin in blood clots, and in keratin in hair. The helical cables in these proteins serve a mechanical role in forming stiff bundles of fibres, as in porcupine quills. The cytoskeleton (internal scaffolding) of cells is rich in so-called intermediate filaments, which also are two-stranded α-helical coiled coils. Many proteins that span biological membranes also contain α helices.

Beta Sheets

Pauling and Corey discovered another periodic structural motif, which they named the *β pleated sheet* (β because it was the second structure that they elucidated, the *α* helix having been the first).

The β pleated sheet (or, more simply, the β sheet) differs markedly from the rodlike *α* helix. A polypeptide chain, called a *β strand,* in a β sheet is almost fully extended rather than being tightly coiled as in the α helix. A range of extended structures are sterically allowed.

The distance between adjacent amino acids along a β strand is approximately 3.5 Å, in contrast with a distance of 1.5 Å along an α helix. The side chains of adjacent amino acids point in opposite directions. A β sheet is formed by linking two or more β strands by hydrogen bonds. Adjacent chains in a β sheet can run in opposite directions (antiparallel β sheet) or in the same direction (parallel β sheet). In the antiparallel arrangement, the NH group and the CO group of each amino acid are respectively hydrogen bonded to the CO group and the NH group of a partner on the adjacent chain. In the parallel arrangement, the hydrogen-bonding scheme is slightly more complicated. For each amino acid, the NH group is hydrogen bonded to the CO group of one amino acid on the adjacent strand, whereas the CO group is hydrogen bonded to the NH group on the amino acid two residues farther along the chain. Many strands, typically 4 or 5 but as many as 10 or more, can come together in β sheets. Such β sheets can be purely antiparallel, purely parallel, or mixed.

In schematic diagrams, β strands are usually depicted by broad arrows pointing in the direction of the carboxyl-terminal end to indicate the type of β sheet formed—parallel or antiparallel. More structurally diverse than α helices, β sheets can be relatively flat but most adopt a somewhat twisted shape. The β sheet is an important structural element in many proteins. For example, fatty acid-binding proteins, important for lipid metabolism, are built almost entirely from β sheets.

Polypeptide Chains

Most proteins have compact, globular shapes, requiring reversals in the direction of their polypeptide chains. Many of these reversals are accomplished by a common structural element called the *reverse turn* (also known as the *β turn* or *hairpin bend*). In many reverse turns, the CO group of residue i of a polypeptide is hydrogen bonded to the NH group of residue $i + 3$. This interaction stabilizes abrupt changes in direction of the polypeptide chain. In other cases, more elaborate structures are responsible for chain reversals. These structures are called *loops* or sometimes *Ω loops* (omega loops) to suggest their overall shape. Unlike α helices and β strands, loops do not have regular, periodic structures. Nonetheless, loop structures are often rigid and well defined. Turns and loops invariably lie on the surfaces of proteins and thus often participate in interactions between proteins and other molecules. The distribution of α helices, β strands, and turns along a protein chain is often referred to as its *secondary structure.*

TERTIARY STRUCTURE

Water-Soluble proteins fold into compact structures with nonpolar cores let us now examine how amino acids are grouped together in a complete protein. X-ray crystallographic and nuclear magnetic resonance studies have revealed the detailed three-dimensional structures of thousands of proteins.

Myoglobin, the oxygen carrier in muscle, is a single polypeptide chain of 153 amino acids. The capacity of myoglobin to bind oxygen depends on the presence of *heme,* a nonpolypeptide *prosthetic (helper) group* consisting of protoporphyrin IX and a central iron atom. *Myo-globin is an extremely compact molecule.* Its overall dimensions are 45 × 35 × 25 Å, an order of magnitude less than if it were fully stretched out. About 70per cent of the main chain is folded into eight α helices, and much of the rest of the chain forms turns and loops between helices.

The folding of the main chain of myoglobin, like that of

most other proteins, is complex and devoid of symmetry. The overall course of the polypeptide chain of a protein is referred to as its *tertiary structure.* A unifying principle emerges from the distribution of side chains. The striking fact is that *the interior consists almost entirely of nonpolar residues* such as leucine, valine, methionine, and phenylalanine. Charged residues such as aspartate, glutamate, lysine, and arginine are absent from the inside of myoglobin. The only polar residues inside are two histidine residues, which play critical roles in binding iron and oxygen. The outside of myoglobin, on the other hand, consists of both polar and nonpolar residues. The spacefilling model shows that there is very little empty space inside.

This contrasting distribution of polar and nonpolar residues reveals a key facet of protein architecture. In an aqueous environment, protein folding is driven by the strong tendency of hydrophobic residues to be excluded from water. Recall that a system is more thermodynamically stable when hydrophobic groups are clustered rather than extended into the aqueous surroundings.

The polypeptide chain therefore folds so that its hydrophobic side chains are buried and its polar, charged chains are on the surface. Many α helices and β strands are amphipathic; that is, the α helix or β strand has a hydrophobic face, which points into the protein interior, and a more polar face, which points into solution.

The fate of the main chain accompanying the hydrophobic side chains is important, too. An unpaired peptide NH or CO group markedly prefers water to a nonpolar milieu.

The secret of burying a segment of main chain in a hydrophobic environment is pairing all the NH and CO groups by hydrogen bonding. This pairing is neatly accomplished in an α helix or β sheet. Van der Waals interactions between tightly packed hydrocarbon side chains also contribute to the stability of proteins. We can now understand why the set of 20 amino acids contains several that differ subtly in size and shape. They provide a palette from which to choose to fill the interior of a protein neatly and thereby maximize van der Waals interactions, which require intimate contact.

Some proteins that span biological membranes are "the exceptions that prove the rule" regarding the distribution of hydrophobic and hydrophilic amino acids throughout three-dimensional structures. For example, consider porins, proteins found in the outer membranes of many bacteria. The permeability barriers of membranes are built largely of alkane chains that are quite hydrophobic. Thus, porins are covered on the outside largely with hydrophobic residues that interact with the neighbouring alkane chains.

In contrast, the centre of the protein contains many charged and polar amino acids that surround a water-filled channel running through the middle of the protein. Thus, because porins function in hydrophobic environments, they are "inside out" relative to proteins that function in aqueous solution.

Some polypeptide chains fold into two or more compact regions that may be connected by a flexible segment of polypeptide chain, rather like pearls on a string. These compact globular units, called *domains,* range in size from about 30 to 400 amino acid residues. For example, the extracellular part of CD4, the cell-surface protein on certain cells of the immune system to which the human immunodeficiency virus (HIV) attaches itself, comprises four similar domains of approximately 100 amino acids each. Often, proteins are found to have domains in common even if their overall tertiary structures are different.

QUATERNARY STRUCTURE

Polypeptide chains can assemble into multisubunit Structures. Four levels of structure are frequently cited in discussions of protein architecture. So far, we have considered three of them.

Primary structure is the amino acid sequence. *Secondary structure* refers to the spatial arrangement of amino acid residues that are nearby in the sequence. Some of these arrangements are of a regular kind, giving rise to a periodic structure.

The α helix and β strand are elements of secondary structure. *Tertiary structure* refers to the spatial arrangement of amino acid residues that are far apart in the sequence and to the pattern of disulfide bonds.

We now turn to proteins containing more than one polypeptide chain. Such proteins exhibit a fourth level of structural organization. Each polypeptide chain in such a protein is called a *subunit. Quaternary structure* refers to the spatial arrangement of subunits and the nature of their interactions. The simplest sort of quaternary structure is a *dimer,* consisting of two identical subunits.

This organization is present in the DNA-binding protein Cro found in a bacterial virus called λ. More complicated quaternary structures also are common. More than one type of subunit can be present, often in variable numbers. For example, human hemoglobin, the oxygen-carrying protein in blood, consists of two subunits of one type (designated α) and two subunits of another type (designated β). Thus, the hemoglobin molecule exists as an $\alpha_2\beta_2$ tetramer. Subtle changes in the arrangement of subunits within the hemoglobin molecule allow it to carry oxygen from the lungs to tissues with great efficiency.

Viruses make the most of a limited amount of genetic information by forming coats that use the same kind of subunit repetitively in a symmetric array. The coat of rhinovirus, the virus that causes the common cold, includes 60 copies each of four subunits. The subunits come together to form a nearly spherical shell that encloses the viral genome.

THREE DIMENSIONAL STRUCTURE OF AMINO ACID

How is the elaborate three-dimensional structure of proteins attained, and how is the three-dimensional structure related to the one-dimensional amino acid sequence information?

The classic work of Christian Anfinsen in the 1950s on the enzyme ribonuclease revealed the relation between the amino acid sequence of a protein and its conformation. Ribonuclease is a single polypeptide chain consisting of 124 amino acid residues cross-linked by four disulfide bonds. Anfinsen's plan was to destroy the three-dimensional structure of the enzyme and to then determine what conditions were required to restore the structure.

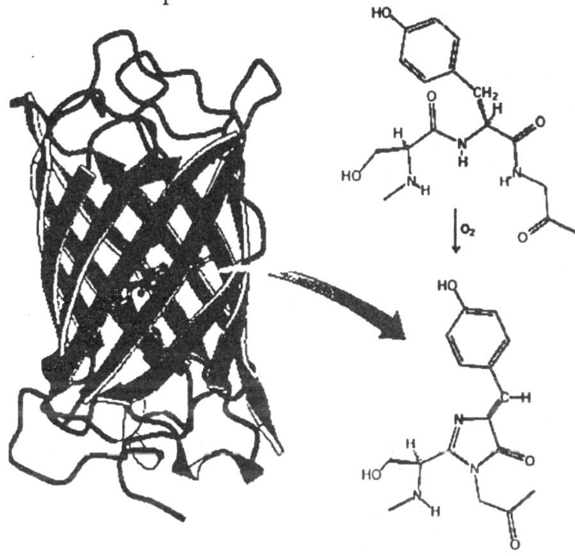

Fig. Chemical Rearrangement in GFP

Agents such as urea or guanidinium chloride effectively disrupt the noncovalent bonds, although the mechanism of action of these agents is not fully understood. The disulfide bonds can be cleaved reversibly by reducing them with a reagent such as *β-mercaptoethanol.* In the presence of a large excess of β-mercaptoethanol, a protein is produced in which the disulfides (cystines) are fully converted into sulfhydryls (cysteines).

HO—CH_2—CH_2—S—H

β-Mercaptoethanol

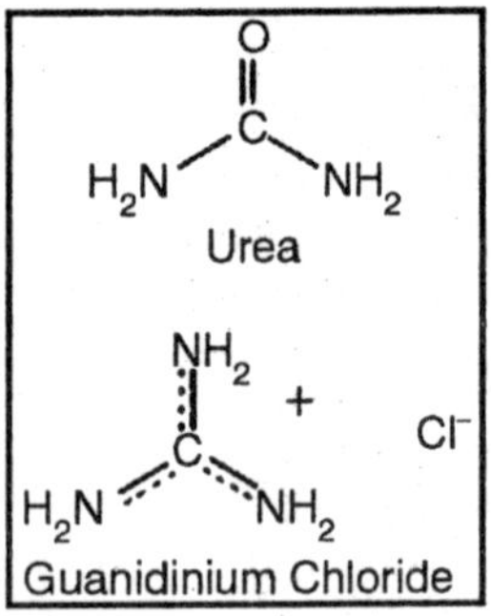

Urea

Guanidinium Chloride

Most polypeptide chains devoid of cross-links assume a *random-coil conformation* in 8 M urea or 6 M guanidinium chloride, as evidenced by physical properties such as viscosity and optical activity. When ribonuclease was treated with β-mercaptoethanol in 8 M urea, the product was a fully reduced, randomly coiled polypeptide chain *devoid of enzymatic activity*. In other words, ribonuclease was *denatured* by this treatment.

Anfinsen then made the critical observation that the denatured ribonuclease, freed of urea and β-mercaptoethanol by dialysis, slowly regained enzymatic activity. He immediately perceived the significance of this chance finding: the sulfhydryl groups of the denatured enzyme became oxidized by air, and the enzyme spontaneously refolded into a catalytically active form.

Detailed studies then showed that nearly all the original enzymatic activity was regained if the sulfhydryl groups were oxidized under suitable conditions. All the measured physical and chemical properties of the refolded enzyme were virtually identical with those of the native enzyme. These experiments showed that *the information needed to specify the catalytically active structure of ribonuclease is contained in its amino acid sequence.* Subsequent studies have established the generality of this central principle of biochemistry: *sequence specifies conformation.* The dependence of conformation on sequence is especially significant because of the intimate connection between conformation and function.

A quite different result was obtained when reduced ribonuclease was reoxidized while it was still in 8 M urea and the preparation was then dialyzed to remove the urea.

Ribonuclease reoxidized in this way had only 1per cent of the enzymatic activity of the native protein. Why were the outcomes so different when reduced ribonuclease was reoxidized in the presence and absence of urea? The reason is that the wrong disulfides formed pairs in urea. There are 105 different ways of pairing eight cysteine molecules to form four disulfides; only one of these combinations is enzymatically active. The 104 wrong pairings have been picturesquely termed "scrambled" ribonuclease.

Anfinsen found that scrambled ribonuclease spontaneously converted into fully active, native ribonuclease when trace amounts of β-mercaptoethanol were added to an aqueous solution of the protein. The added β-mercaptoethanol catalyzed the rearrangement of disulfide pairings until the native structure was regained in about 10 hours. *This process was driven by the decrease in free energy as the scrambled conformations were converted into the stable, native conformation of the enzyme.* The native disulfide pairings of ribonuclease thus contribute to the stabilization of the thermodynamically preferred structure.

Similar refolding experiments have been performed on many other proteins. In many cases, the native structure can be generated under suitable conditions. For other proteins, however, refolding does not proceed efficiently. In these cases, the unfolding protein molecules usually become tangled up with one another to form aggregates. Inside cells, proteins called *chaperones* block such illicit interactions.

Amino Acids Propensities

How does the amino acid sequence of a protein specify its three-dimensional structure? How does an unfolded polypeptide chain acquire the form of the native protein? These fundamental questions in biochemistry can be approached by first asking a simpler one: What determines whether a particular sequence in a protein forms an α helix, a β strand, or a turn?

Examining the frequency of occurrence of particular amino acid residues in these secondary structures can be a source of insight into this determination. Residues such as alanine, glutamate, and leucine tend to be present in α helices, whereas valine and isoleucine tend to be present in β strands. Glycine, asparagine, and proline have a propensity for being in turns.

The results of studies of proteins and synthetic peptides have revealed some reasons for these preferences. The α helix can be regarded as the default conformation. Branching at the β-carbon atom, as in valine, threonine, and isoleucine, tends to destabilize *α* helices because of steric clashes. These residues are readily accommodated in β strands, in which their side chains project out of the plane containing the main chain. Serine, aspartate, and asparagine tend to disrupt α helices because their side chains contain hydrogen-bond donors or acceptors in close proximity to the main chain, where they compete for main-chain NH and CO groups.

Proline tends to disrupt both α helices and β strands because it lacks an NH group and because its ring structure restricts its φ value to near -60 degrees. Glycine readily fits into all structures and for that reason does not favour helix formation in particular.

Can one predict the secondary structure of proteins by using this knowledge of the conformational preferences of amino acid residues? Predictions of secondary structure adopted by a stretch of six or fewer residues have proved to be about 60 to 70per cent accurate. What stands in the way of more accurate prediction? Note that the conformational preferences of amino acid residues are not tipped all the way to one structure.

For example, glutamate, one of the strongest helix formers, prefers α helix to β strand by only a factor of two. The preference ratios of most other residues are smaller. Indeed, some penta- and hexapeptide sequences have been found to adopt one structure in one protein and an entirely different structure in another. Hence, some amino acid sequences do not uniquely determine secondary structure. Tertiary interactions—interactions between residues that are far apart in the sequence—may be decisive in specifying the secondary structure of some segments. *The context is often crucial in determining the conformational outcome.* The conformation of a protein evolved to work in a particular environment or context.

Pathological conditions can result if a protein assumes an inappropriate conformation for the context. Striking examples are *prion diseases,* such as Creutzfeldt-Jacob disease, kuru, and mad cow disease. These conditions result when a brain protein called a prion converts from its normal conformation (designated PrP^{C}) to an altered one (PrP^{Sc}). This conversion is self-propagating, leading to large aggregates of PrP^{Sc}. The role of these aggregates in the generation of the pathological conditions is not yet understood.

Protein Folding

As stated earlier, proteins can be denatured by heat or by chemical denaturants such as urea or guanidium chloride. For many proteins, a comparison of the degree of unfolding as the concentration of denaturant increases has revealed a relatively sharp transition from the folded, or native, form to the unfolded, or denatured, form, suggesting that only these two conformational states are present to any significant extent. A similar sharp transition is observed if one starts with unfolded proteins and removes the denaturants, allowing the proteins to fold.

Protein folding and unfolding is thus largely an *"all or none" process* that results from a *cooperative transition.* For example, suppose that a protein is placed in conditions under which some part of the protein structure is thermodynamically unstable. As this part of the folded structure is disrupted, the interactions between it and the remainder of the protein will be lost. The loss of these interactions, in turn, will destabilize the remainder of the structure. Thus, conditions that lead to the disruption of any part of a protein structure are likely to unravel the protein completely. The structural properties of proteins provide a clear rationale for the cooperative transition.

The consequences of cooperative folding can be

illustrated by considering the contents of a protein solution under conditions corresponding to the middle of the transition between the folded and unfolded forms. Under these conditions, the protein is "half folded." Yet the solution will contain no half-folded molecules but, instead, will be a 50/50 mixture of fully folded and fully unfolded molecules. Structures that are partly intact and partly disrupted are not thermodynamically stable and exist only transiently. Cooperative folding ensures that partly folded structures that might interfere with processes within cells do not accumulate.

Proteins Folding by Stabilization

The cooperative folding of proteins is a thermodynamic property; its occurrence reveals nothing about the kinetics and mechanism of protein folding. How does a protein make the transition from a diverse ensemble of unfolded structures into a unique conformation in the native form? One possibility a priori would be that all possible conformations are tried out to find the energetically most favourable one. How long would such a random search take? Consider a small protein with 100 residues. Cyrus Levinthal calculated that, if each residue can assume three different conformations, the total number of structures would be 3^{100}, which is equal to 5×10^{47}. If it takes 10^{-13} s to convert one structure into another, the total search time would be $5 \times 10^{47} \times 10^{-13}$ s, which is equal to 5×10^{34} s, or 1.6×10^{27} years. Clearly, it would take much too long for even a small protein to fold properly by randomly trying out all possible conformations. The enormous difference between calculated and actual folding times is called *Levinthal's paradox.*

The way out of this dilemma is to recognize the power of *cumulative selection*. Richard Dawkins, in *The Blind Watchmaker,* asked how long it would take a monkey poking randomly at a typewriter to reproduce Hamlet's remark to Polonius, "Methinks it is like a weasel". An astronomically large number of keystrokes, of the order of 10^{40}, would be required. However, suppose that we preserved each correct character and allowed the monkey to retype only the wrong ones. In this case, only a few thousand keystrokes, on average, would be needed. The crucial difference between these cases is that the first employs a completely random search, whereas, in the second, *partly correct intermediates are retained.*

The essence of protein folding is the retention of partly correct intermediates. However, the protein-folding problem is much more difficult than the one presented to our simian Shakespeare. First, the criterion of correctness is not a residue-by-residue scrutiny of conformation by an omniscient observer but rather the total free energy of the transient species. Second, proteins are only marginally stable. The free-energy difference between the folded and the unfolded states of a typical 100-residue protein is 10 kcal mol^{-1} (42 kJ mol^{-1}), and thus each residue contributes on average only 0.1 kcal mol^{-1} (0.42 kJ mol^{-1}) of energy to maintain the folded state. This amount is less than that of thermal energy, which is 0.6 kcal mol^{-1} (2.5 kJ mol^{-1}) at room temperature. This meager stabilization energy means that correct intermediates, especially those formed early in folding, can be lost. The analogy is that the monkey would be somewhat free to undo its correct keystrokes.

Nonetheless, the interactions that lead to cooperative folding can stabilize intermediates as structure builds up. Thus, local regions, which have significant structural preference, though not necessarily stable on their own, will tend to adopt their favoured structures and, as they form, can interact with one other, leading to increasing stabilization.

Prediction of Structure

The amino acid sequence completely determines the three-dimensional structure of a protein. However, the prediction of three-dimensional structure from sequence has proved to be extremely difficult. As we have seen, the local sequence appears to determine only between 60per cent and 70per cent of the secondary structure; long-range interactions are required to fix the full secondary structure and the tertiary structure.

Investigators are exploring two fundamentally different approaches to predicting three-dimensional structure from amino acid sequence.

The first is *ab initio prediction,* which attempts to predict the folding of an amino acid sequence without any direct reference to other known protein structures. Computer-based calculations are employed that attempt to minimize the free energy of a structure with a given amino acid sequence or to simulate the folding process.

The utility of these methods is limited by the vast number of possible conformations, the marginal stability of proteins, and the subtle energetics of weak interactions in aqueous solution. The second approach takes advantage of our growing knowledge of the three-dimensional structures of many proteins. In these *knowledge-based methods,* an amino acid sequence of unknown structure is examined for compatibility with any known protein structures. If a significant match is detected, the known structure can be used as an initial model. Knowledge-based methods have been a source of many insights into the three-dimensional conformation of proteins of known sequence but unknown structure.

Protein Modification

Proteins are able to perform numerous functions relying solely on the versatility of their 20 amino acids. However, many proteins are covalently modifed, through the attachment of groups other than amino acids, to augment their functions. For example, *acetyl groups* are attached to the amino termini of many proteins, a modification that makes these proteins more resistant to degradation.

The addition of *hy-droxyl groups* to many proline residues stabilizes fibres of newly synthesized collagen, a fibrous protein found in connective tissue and bone. The biological significance of this modification is evident in the disease scurvy: a deficiency of vitamin C results in insufficient hydroxylation of collagen and the abnormal collagen fibres that result are unable to maintain normal tissue strength.

Another specialized amino acid produced by a finishing touch is *γ-carboxyglutamate*. In vitamin K deficiency, insufficient carboxylation of glutamate in prothrombin, a clotting protein, can lead to hemorrhage.

Many proteins, especially those that are present on the surfaces of cells or are secreted, acquire *carbohydrate units* on specific asparagine residues. The addition of sugars makes the proteins more hydrophilic and able to participate in interactions with other proteins. Conversely, the addition of a *fatty acid* to an α-amino group or a cysteine sulfhydryl group produces a more hydrophobic protein.

Many hormones, such as epinephrine (adrenaline), alter the activities of enzymes by stimulating the phosphorylation of the hydroxyl amino acids serine and threonine; *phosphoserine* and *phosphothreonine* are the most ubiquitous modified amino acids in proteins. Growth factors such as insulin act by triggering the phosphorylation of the hydroxyl group of tyrosine residues to form *phosphotyrosine*. The phosphoryl groups on these three modified amino acids are readily removed; thus they are able to act as reversible switches in regulating cellular processes.

The preceding modifications consist of the addition of special groups to amino acids. Other special groups are generated by chemical rearrangements of side chains and, sometimes, the peptide backbone. For example, certain jellyfish produce a fluorescent green protein. The source of the fluorescence is a group formed by the spontaneous rearrangement and oxidation of the sequence Ser-Tyr-Gly within the centre of the protein. This protein is of great utility to researchers as a marker within cells.

Finally, many proteins are cleaved and trimmed after synthesis. For example, digestive enzymes are synthesized as inactive precursors that can be stored safely in the pancreas. After release into the intestine, these precursors become activated by peptide-bond cleavage.

In blood clotting, peptide-bond cleavage converts soluble fibrinogen into insoluble fibrin. A number of polypeptide hormones, such as adrenocorticotropic hormone, arise from the splitting of a single large precursor protein. Likewise, many virus proteins are produced by the cleavage of large polyprotein precursors. We shall encounter many more examples of modification and cleavage as essential features of protein formation and function. Indeed, these finishing touches account for much of the versatility, precision, and elegance of protein action and regulation.

Chapter 5

Carbohydrates

INTRODUCTION

Carbohydrates are the most abundant class of organic compounds found in living organisms. They originate as products of photosynthesis, an endothermic reductive condensation of carbon dioxide requiring light energy and the pigment chlorophyll.

$$n\ CO_2 + n\ H_2O + \text{energy}\ C_nH_{2n}O_n + n\ O_2$$

Complexity Carbohydrates	**Simple Carbohydrates**	**Complex**
	monosaccharides	disaccharides, oligosaccharides & polysaccharides

Size	**Tetrose**	**Pentose**	**Hexose**	**Heptose**	
	C_4 sugars	C_5 sugars	C_6 sugars	C_7 sugars	etc.

C=O Function	**Aldose** sugars having an aldehyde function or an acetal equivalent. **Ketose** sugars having a ketone function or an acetal equivalent.
Reactivity	**Reducing** sugars oxidized by Tollens' reagent (or Benedict's or Fehling's reagents). **Non-reducing** sugars not oxidized by Tollens' or other reagents.

As noted here, the formulas of many carbohydrates can be written as carbon hydrates, $C_n(H_2O)_n$, hence their name. The carbohydrates are a major source of metabolic energy, both for plants and for animals that depend on plants for food. Aside from the sugars and starches that meet this vital nutritional role, carbohydrates also serve as a structural material (cellulose), a component of the energy transport compound ATP, recognition sites on cell surfaces and one of three essential components of DNA and RNA.

Carbohydrates are called saccharides or, if they are relatively small, sugars. Several classifications of carbohydrates have proven useful and are outlined in the above table.

GLUCOSE

Carbohydrates have been given non-systematic names, although the suffix ose is generally used. The most common carbohydrate is glucose ($C_6H_{12}O_6$). Glucose is a monosaccharide, an aldohexose and a reducing sugar. The general structure of glucose and many other aldohexoses was established by simple chemical reactions. The following diagram illustrates the kind of evidence considered, although some of the reagents shown here are different from those used by the original scientists.

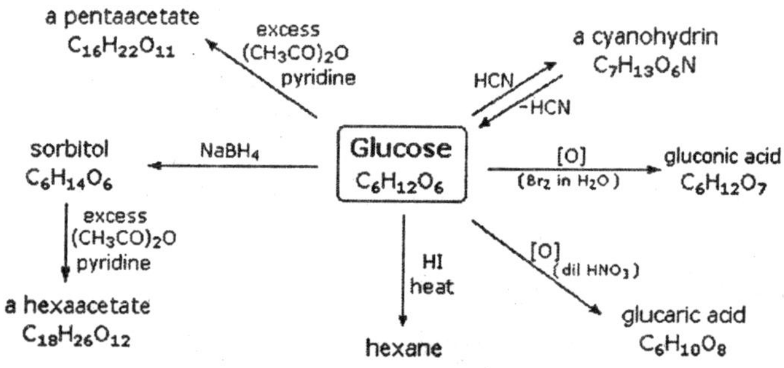

Fig. Glucose

Hot hydriodic acid (HI) was often used to reductively remove oxygen functional groups from a molecule and in the case of glucose this treatment gave hexane (in low yield). From this it was concluded that the six carbons are in an unbranched chain. The presence of an aldehyde carbonyl group was deduced from cyanohydrin formation, its reduction to the hexa-alcohol sorbitol, also called glucitol and mild oxidation to the mono-carboxylic acid, glucuronic acid. Somewhat stronger oxidation by dilute nitric acid gave the diacid, glucaric acid, supporting the proposal of a six-carbon chain.

The five oxygens remaining in glucose after the aldehyde was accounted for were thought to be in hydroxyl groups, since a penta-acetate derivative could be made. These hydroxyl groups were assigned, one each, to the last five carbon atoms, because geminal hydroxyl groups are normally unstable relative to the carbonyl compound formed by loss of water. The four middle carbon atoms in the glucose chain are centres of chirality and are coloured red.

Glucose and other saccharides are extensively cleaved by periodic acid, thanks to the abundance of vicinal diol moieties in their structure. This oxidative cleavage, known as the Malaprade reaction is particularly useful for the analysis of selective O-substituted derivatives of saccharides, since ether functions do not react. The stoichiometry of aldohexose cleavage is shown in the following equation.

$$HOCH_2(CHOH)_4CHO + 5\ HIO_4 \rightarrow H_2C = O + 5\ HCO_2H + 5\ HIO_3$$

CONFIGURATION OF GLUCOSE

The four chiral centres in glucose indicate there may be as many as sixteen (2^4) stereoisomers having this constitution. These would exist as eight diastereomeric pairs of enantiomers and the initial challenge was to determine which of the eight corresponded to glucose. This challenge was accepted and met in 1891 by the German chemist Emil Fischer.

His successful negotiation of the stereochemical maze presented by the aldohexoses was a logical tour de force and it is fitting that he received the 1902 Nobel Prize for chemistry for this accomplishment. One of the first tasks faced by Fischer was to devise a method of representing the configuration of each chiral cenre in an unambiguous manner. To this end, he invented a simple technique for drawing chains of chiral centres, that we now call the Fischer projection formula. At the time Fischer undertook the glucose project it was not possible to establish the absolute configuration of an enantiomer. Consequently, Fischer made an arbitrary choice for (+)-glucose and established a network of related aldose configurations that he called the D-family.

The mirror images of these configurations were then designated the L-family of aldoses. To illustrate using present day knowledge, Fischer projection formulas and names for the D-aldose family (three to six-carbon atoms) are shown below, with the asymmetric carbon atoms (chiral centres) coloured red.

The last chiral cenre in an aldose chain (farthest from the aldehyde group) was chosen by Fischer as the D/L designator site. If the hydroxyl group in the projection formula pointed to the right, it was defined as a member of the D-family. A left directed hydroxyl group then represented the L-family.

Fischer's initial assignment of the D-configuration had a 50:50 chance of being right, but all his subsequent conclusions concerning the relative configurations of various aldoses were soundly based. In 1951 x-ray fluorescence studies of (+)-tartaric acid, carried out in the Netherlands by Johannes Martin Bijvoet (pronounced "buy foot"), proved that Fischer's choice was correct. It is important to recognize that the sign of a compound's specific rotation (an experimental number) does not correlate with its configuration (D or L). It is a simple matter to measure an optical rotation with a polarimetre. Determining an absolute configuration usually requires chemical interconversion with known compounds by stereospecific reaction paths.

OXIDATION

Sugars may be classified as reducing or non-reducing

based on their reactivity with Tollens', Benedict's or Fehling's reagents. If a sugar is oxidized by these reagents it is called reducing, since the oxidant (Ag(+) or Cu(+2)) is reduced in the reaction, as evidenced by formation of a silver mirror or precipitation of cuprous oxide. The Tollens' test is commonly used to detect aldehyde functions; and because of the facile interconversion of ketoses and aldoses under the basic conditions of this test, ketoses such as fructose also react and are classified as reducing sugars.

When the aldehyde function of an aldose is oxidized to a carboxylic acid the product is called an aldonic acid. Because of the 2º hydroxyl functions that are also present in these compounds, a mild oxidizing agent such as hypobromite must be used for this conversion.

If both ends of an aldose chain are oxidized to carboxylic acids the product is called an aldaric acid. By converting an aldose to its corresponding aldaric acid derivative, the ends of the chain become identical (this could also be accomplished by reducing the aldehyde to CH_2OH, as noted below). Such an operation will disclose any latent symmetry in the remaining molecule. Thus, ribose, xylose, allose and galactose yield achiral aldaric acids which are, of course, not optically active. Other aldose sugars may give identical chiral aldaric acid products, implying a unique configurational relationship. Remember, a Fischer projection formula may be rotated by 180º in the plane of projection without changing its configuration.

REDUCTION

Sodium borohydride reduction of an aldose makes the ends of the resulting alditol chain identical, $HOCH_2(CHOH)_nCH_2OH$, thereby accomplishing the same configurational change produced by oxidation to an aldaric acid. Thus, allitol and galactitol from reduction of allose and galactose are achiral and altrose and talose are reduced to the same chiral alditol. A summary of these redox reactions and derivative nomenclature is given in the following table.

Derivatives of $HOCH_2(CHOH)_nCHO$

HOBr Oxidation → $HOCH_2(CHOH)_nCO_2H$ an Aldonic Acid

HNO_3 Oxidation → $H_2OC(CHOH)_nCO_2H$ an Aldaric Acid

$NaBH_4$ Reduction → $HOCH_2(CHOH)_nCH_2OH$ an Alditol

OSAZONE FORMATION

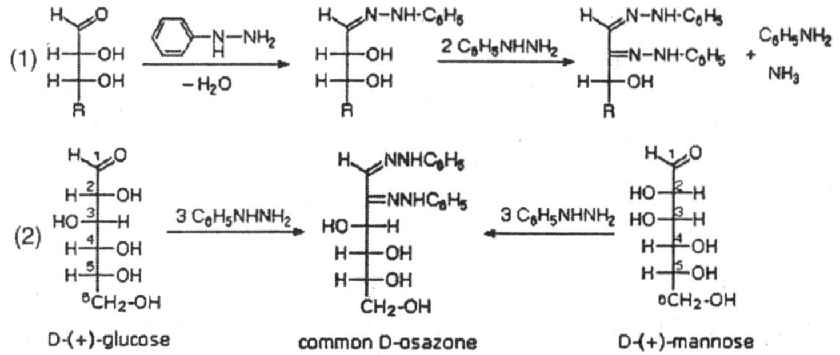

(1) H–C=O / H–C–OH / H–C–OH / R —[$C_6H_5NHNH_2$, $-H_2O$]→ H–C=N–NH–C_6H_5 / H–C–OH / H–C–OH / R —[2 $C_6H_5NHNH_2$]→ H–C=N–NH–C_6H_5 / C=N–NH–C_6H_5 / H–C–OH / R + $C_6H_5NH_2$ + NH_3

(2) D-(+)-glucose —[3 $C_6H_5NHNH_2$]→ common D-osazone ←[3 $C_6H_5NHNH_2$]— D-(+)-mannose

The osazone reaction was developed and used by Emil Fischer to identify aldose sugars differing in configuration only at the alpha-carbon.

The upper equation shows the general form of the osazone reaction, which effects an alpha-carbon oxidation with formation of a bis-phenylhydrazone, known as an osazone. Application of the osazone reaction to D-glucose and D-mannose demonstrates that these compounds differ in configuration only at C-2.

CHAIN SHORTENING AND LENGTHENING

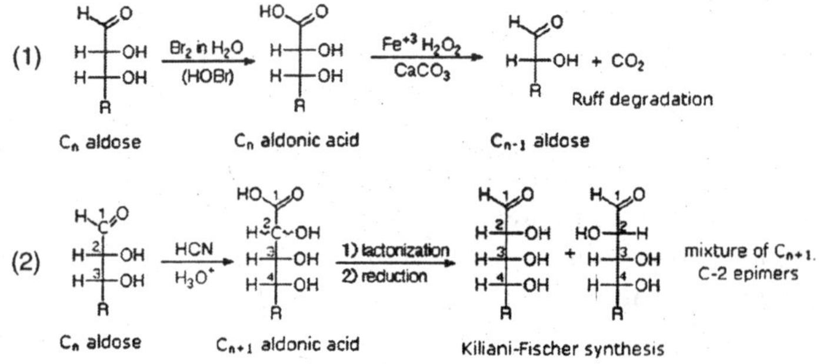

(1) C_n aldose —[Br_2 in H_2O (HOBr)]→ C_n aldonic acid —[Fe^{+3} H_2O_2, $CaCO_3$]→ C_{n-1} aldose + CO_2

Ruff degradation

(2) C_n aldose —[HCN, H_3O^+]→ C_{n+1} aldonic acid —[1) lactonization 2) reduction]→ mixture of C_{n+1} C-2 epimers

Kiliani-Fischer synthesis

These two procedures permit an aldose of a given size to be related to homologous smaller and larger aldoses. The importance of these relationships may be seen in the array of aldose structures presented earlier, where the structural connections are given by the dashed blue lines. Thus Ruff degradation of the pentose arabinose gives the tetrose erythrose. Working in the opposite direction, a Kiliani-Fischer synthesis applied to arabinose gives a mixture of glucose and mannose. Using these reactions we can now follow Fischer's train of logic in assigning the configuration of D-glucose.

- Ribose and arabinose (two well known pentoses) both gave erythrose on Ruff degradation.

  As expected, Kiliani-Fischer synthesis applied to erythrose gave a mixture of ribose and arabinose.
- Oxidation of erythrose gave an achiral (optically inactive) aldaric acid. This defines the configuration of erythrose.
- Oxidation of ribose gave an achiral (optically inactive) aldaric acid. This defines the configuration of both ribose and arabinose.
- Ruff shortening of glucose gave arabinose and Kiliani-Fischer synthesis applied to arabinose gave a mixture of glucose and mannose.
- Glucose and mannose are therefore epimers at C-2, a fact confirmed by the common product from their osazone reactions.
- A pair of structures for these epimers can be written, but which is glucose and which is mannose?

In order to determine which of these epimers was glucose, Fischer made use of the inherent C_2 symmetry in the four-carbon dissymmetric core of one epimer (B). This is shown in the following diagram by a red dot where the symmetry axis passes through the projection formula. Because of this symmetry, if the aldehyde and 1º-alcohol functions at the ends of the chain are exchanged, epimer B would be unchanged; whereas A would be converted to a different compound.

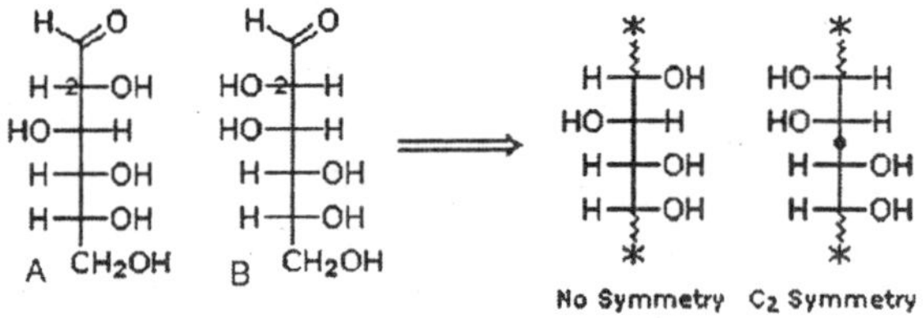

A B

No Symmetry C_2 Symmetry

Fischer looked for and discovered a second aldohexose that represented the end group exchange for the epimer lacking the latent C_2 symmetry (A). This compound was L-(+)-gulose and its exchange relationship to D-(+)-glucose was demonstrated be oxidation to a common aldaric acid product.

KETOSES

If a monosaccharide has a carbonyl function on one of the inner atoms of the carbon chain it is classified as a ketose. Dihydroxyacetone may not be a sugar, but it is included as the ketose analog of glyceraldehyde. The carbonyl group is commonly found at C-2, as illustrated by the following examples (chiral centres are coloured red). As expected, the carbonyl function of a ketose may be reduced by sodium borohydride, usually to a mixture of epimeric products. D-Fructose, the sweetest of the common natural sugars, is for example reduced to a mixture of D-glucitol (sorbitol) and D-mannitol, named after the aldohexoses from which they may also be obtained by analogous reduction. Mannitol is itself a common natural carbohydrate. Although the ketoses are distinct isomers of the aldose monosaccharides, the chemistry of both classes is linked due to their facile interconversion in the presence of acid or base catalysts. This interconversion and the corresponding epimerization at sites alpha to the carbonyl functions, occurs by way of an enediol tautomeric intermediate.

Ketose Examples

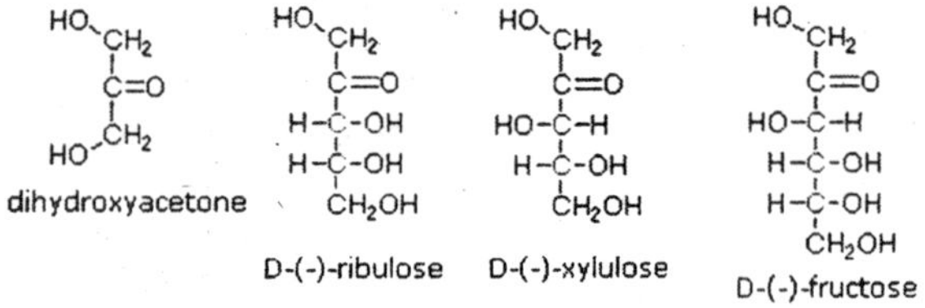

dihydroxyacetone D-(-)-ribulose D-(-)-xylulose D-(-)-fructose

Because of base-catalyzed isomerizations of this kind, the Tollens' reagent is not useful for distinguishing aldoses from ketoses or for specific oxidation of aldoses to the corresponding aldonic acids. Oxidation by HOBr is preferred for the latter conversion.

ANOMERIC FORMS OF GLUCOSE

Fischers brilliant elucidation of the configuration of glucose did not remove all uncertainty concerning its structure. Two different crystalline forms of glucose were reported in 1895. Each of these gave all the characteristic reactions of glucose and when dissolved in water equilibrated to the same mixture. This equilibration takes place over a period of many minutes and the change in optical activity that occurs is called mutarotation. These facts are summarized in the diagram below.

When glucose was converted to its pentamethyl ether (reaction with excess CH_3I & AgOH), two different isomers were isolated and neither exhibited the expected aldehyde reactions. Acid-catalyzed hydrolysis of the pentamethyl ether derivatives, however, gave a tetramethyl derivative that was oxidized by Tollen's reagent and reduced by sodium borohydride, as expected for an aldehyde.

The search for scientific truth often proceeds in stages and the structural elucidation of glucose serves as a good example. Somehow a new stereogenic cenre must be created and the aldehyde must be deactivated in the pentamethyl derivative. A simple solution to this dilemma is achieved by converting the open aldehyde structure for glucose into a cyclic hemiacetal, called a glucopyranose.

The linear aldehyde is tipped on its side and rotation about the C_4-C_5 bond brings the C5-hydroxyl function close to the aldehyde carbon. For ease of viewing, the six-membered hemiacetal structure is drawn as a flat hexagon, but it actually assumes a chair conformation.

The hemiacetal carbon atom (C_{-1}) becomes a new stereogenic cenre, commonly referred to as the anomeric carbon and the α and β-isomers are called anomers. First, we know that hemiacetals are in equilibrium with their carbonyl and alcohol components when in solution. Consequently, fresh solutions of either alpha or beta-glucose crystals in water should establish an equilibrium mixture of both anomers, plus the open chain chain form.

Second, a pentamethyl ether derivative of the pyranose structure converts the hemiacetal function to an acetal. Acetals are stable to base, so this product should not react with Tollen's reagent or be reduced by sodium borohydride. Acid hydrolysis of acetals regenerates the carbonyl and alcohol components and in the case of the glucose derivative this will be a tetramethyl ether of the pyranose hemiacetal. This compound will, of course, undergo typical aldehyde reactions.

CYCLIC FORMS OF MONOSACCHARIDES

The preferred structural form of many monosaccharides may be that of a cyclic hemiacetal. Five and six-membered rings are favoured over other ring sizes because of their low angle and eclipsing strain.

Cyclic structures of this kind are termed furanose (five-membered) or pyranose (six-membered), reflecting the ring size relationship to the common heterocyclic compounds furan and pyran shown on the right. Ribose, an important aldopentose, commonly adopts a furanose structure by convention for the D-family, the five-membered furanose ring is drawn in an edgewise projection with the ring oxygen positioned away from the viewer.

The anomeric carbon atom (coloured red here) is placed on the right. The upper bond to this carbon is defined as beta, the lower bond then is alpha. The cyclic pyranose forms of various monosaccharides are often drawn in a flat projection known as a Haworth formula, after the British chemist, Norman Haworth.

As with the furanose ring, the anomeric carbon is placed on the right with the ring oxygen to the back of the edgewise view. In the D-family, the alpha and beta bonds have the same orientation defined for the furanose ring. These Haworth formulas are convenient for displaying stereochemical relationships, but do not represent the true shape of the molecules. We know that these molecules are actually puckered in a fashion we call a chair conformation. Examples of four typical pyranose structures are shown below, both as Haworth projections and as the more representative chair conformers. The anomeric carbons are coloured red.

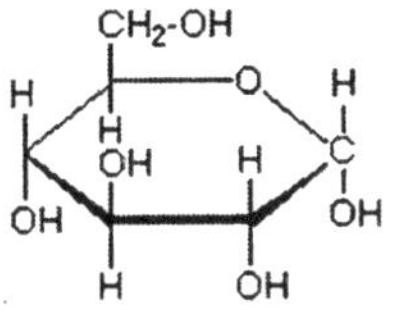

α-D-glucopyranose 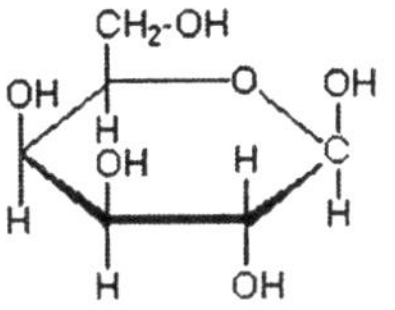 β-D-galactopyranose

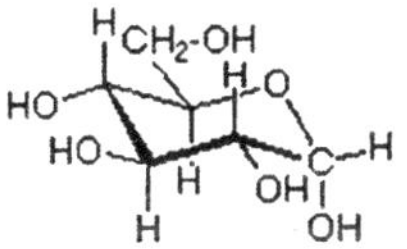

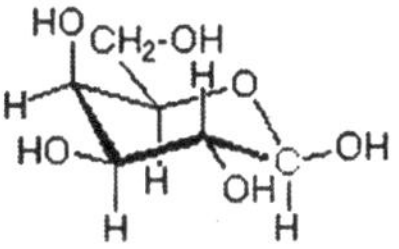

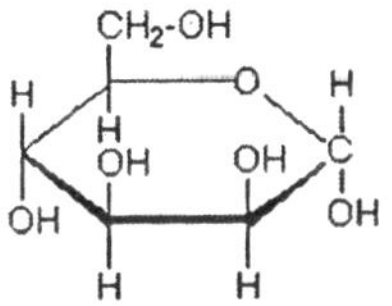

α-D-mannopyranose 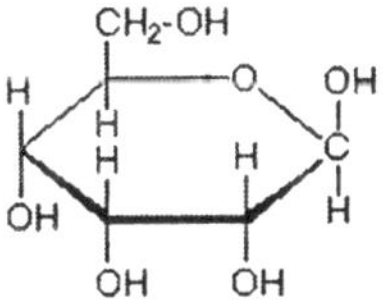 β-D-allopyranose

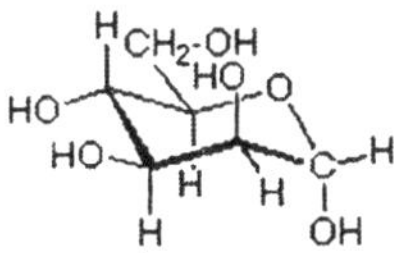

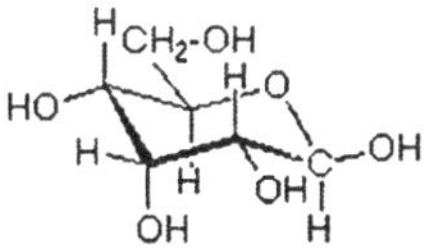

The size of the cyclic hemiacetal ring adopted by a given sugar is not constant, but may vary with substituents and other structural features. Aldolhexoses usually form pyranose rings and their pentose homologs tend to prefer the furanose form, but there are many counter examples. The formation of acetal derivatives illustrates how subtle changes may alter this selectivity. A pyranose structure for D-glucose is drawn in the rose-shaded box on the left. Acetal derivatives have been prepared by acid-catalyzed reactions with benzaldehyde and acetone. As a rule, benzaldehyde forms six-membered cyclic acetals, whereas acetone prefers to form five-membered acetals. The top equation shows the formation and some reactions of the 4,6-O-benzylidene acetal, a commonly employed protective group.

A methyl glycoside derivative of this compound leaves the C-2 and C-3 hydroxyl groups exposed to reactions such as the periodic acid cleavage, shown as the last step. The formation of an isopropylidene acetal at C-1 and C-2, cenre structure, leaves the C-3 hydroxyl as the only unprotected function. Selective oxidation to a ketone is then possible. Finally, direct di-O-isopropylidiene derivatization of glucose by reaction with excess acetone results in a change to a furanose structure in which the C-3 hydroxyl is again unprotected. However, the same reaction with D-galactose, shown in the blue-shaded box, produces a pyranose product in which the C-6 hydroxyl is unprotected. Both derivatives do not react with Tollens' reagent. This difference in behaviour is attributed to the cis-orientation of the C-3 and C-4 hydroxyl groups in galactose, which permits formation of a less strained five-membered cyclic acetal, compared with the trans-C-3 and C-4 hydroxyl groups in glucose. Derivatizations of this kind permit selective reactions to be conducted at different locations in these highly functionalized molecules.

GLYCOSIDES

Acetal derivatives formed when a monosaccharide reacts with an alcohol in the presence of an acid catalyst are called glycosides. This reaction is illustrated for glucose and methanol in the diagram below. In naming of glycosides, the "oseΔ suffix of the sugar name is replaced by "osideΔ and the alcohol group name is placed first. As is generally true for most aldols, glycoside formation involves the loss of an equivalent of water. The diether product is stable to base and alkaline oxidants such as Tollen's reagent. Since acid-catalyzed aldolization is reversible, glycosides may be hydrolyzed back to their alcohol and sugar components by aqueous acid.

The anomeric methyl glucosides are formed in an equilibrium ratio of 66% alpha to 34% beta. In the case of glucose, the substituents on the beta-anomer are all equatorial, whereas the C-1 substituent in the alpha-anomer changes to axial. Since substituents on cyclohexane rings prefer an equatorial location over axial (methoxycyclohexane is 75% equatorial), the preference for alpha-glycopyranoside formation is unexpected and is referred to as the anomeric effect. Glycosides abound in biological systems. By attaching a sugar moiety to a lipid or benzenoid structure, the solubility and other properties of the compound may be changed substantially. Because of the important modifying influence of such derivatization, numerous enzyme systems, known as glycosidases, have evolved for the attachment and removal of sugars from alcohols, phenols and amines. Chemists refer to the sugar component of natural glycosides as the glycon and the alcohol component as the aglycon. Salicin, one of the oldest herbal remedies known, was the model for the synthetic analgesic aspirin.

A large class of hydroxylated, aromatic oxonium cations called anthocyanins provide the red, purple and blue colours of many flowers, fruits and some vegetables. Peonin is one example of this class of natural pigments, which exhibit a pronounced pH colour dependence. The oxonium moiety is only stable in acidic environments and the colour changes or disappears when base is added. The complex changes that occur when wine is fermented and stored are in part associated with glycosides of anthocyanins. Finally, amino derivatives of ribose, such as cytidine play important roles in biological phosphorylating agents, coenzymes and information transport and storage materials.

DISACCHARIDES

When the alcohol component of a glycoside is provided by a hydroxyl function on another monosaccharide, the compound is called a disaccharide. Four examples of disaccharides composed of two glucose units are shown in the following diagram. The individual glucopyranose rings are labled A and B and the glycoside bonding is circled in light blue. Notice that the glycoside bond may be alpha, as in maltose and trehalose, or beta as in cellobiose and gentiobiose. Acid-catalyzed hydrolysis of these disaccharides yields glucose as the only product.

Enzyme-catalyzed hydrolysis is selective for a specific glycoside bond, so an alpha-glycosidase cleaves maltose and trehalose to glucose, but does not cleave cellobiose or gentiobiose. A beta-glycosidase has the opposite activity. In order to draw a representative structure for cellobiose, one of the glucopyranose rings must be rotated by 180º, but this feature is often omitted in favour of retaining the usual perspective for the individual rings.

The bonding between the glucopyranose rings in cellobiose and maltose is from the anomeric carbon in ring A to the C-4 hydroxyl group on ring B. This leaves the anomeric carbon in ring B free, so cellobiose and maltose both may assume alpha and beta anomers at that site . Gentiobiose has a beta-glycoside link, originating at C-1 in ring A and terminating at C-6 in ring B. Its alpha-anomer is drawn in the diagram. Because cellobiose, maltose and gentiobiose are hemiacetals they are all reducing sugars (oxidized by Tollen's reagent). Trehalose, a disaccharide found in certain mushrooms, is a bis-acetal and is therefore a non-reducing sugar. A systematic nomenclature for disaccharides exists, but as the following examples illustrate, these are often lengthy.

Cellobiose: 4-O-β-D-Glucopyranosyl-D-glucose
Maltose: 4-O-α-D-Glucopyranosyl-D-glucose
Gentiobiose: 6-O-β-D-Glucopyranosyl-D-glucose
Trehalose: α-D-Glucopyranosyl-α-D-glucopyranoside

Although all the disaccharides shown here are made up of two glucopyranose rings, their properties differ in interesting ways. Maltose, sometimes called malt sugar, comes from the hydrolysis of starch. It is about one third as sweet as cane sugar (sucrose), is easily digested by humans and is fermented by yeast. Cellobiose is obtained by the hydrolysis of cellulose. It has virtually no taste, is indigestible by humans and is not fermented by yeast. Some bacteria have beta-glucosidase enzymes that hydrolyze the glycosidic bonds in cellobiose and cellulose. The presence of such bacteria in the digestive tracts of cows and termites permits these animals to use cellulose as a food. Finally, it may be noted that trehalose has a distinctly sweet taste, but gentiobiose is bitter.

Disaccharides made up of other sugars are known, but glucose is often one of the components.. Lactose, also known as milk sugar, is a galactose-glucose compound joined as a beta-glycoside. It is a reducing sugar because of the hemiacetal function remaining in the glucose moiety. Many adults, particulaly those from regions where milk is not a dietary staple, have a metabolic intolerance for lactose. Infants have a digestive enzyme which cleaves the beta-glycoside bond in lactose, but production of this enzyme stops with weaning. Cheese is less subject to the lactose intolerance problem, since most of the lactose is removed with the whey. Sucrose, or cane sugar, is our most commonly used sweetening agent. It is a non-reducing disaccharide composed of glucose and fructose joined at the anomeric carbon of each by glycoside bonds (one alpha and one beta). In the formula shown here the fructose ring has been rotated 180º from its conventional perspective.

POLYSACCHARIDES

As the name implies, polysaccharides are large high-molecular weight molecules constructed by joining monosaccharide units together by glycosidic bonds. They are sometimes called glycans. The most important compounds in this class, cellulose, starch and glycogen are all polymers of glucose. This is easily demonstrated by acid-catalyzed hydrolysis to the monosaccharide. Since partial hydrolysis of cellulose gives varying amounts of cellobiose, we conclude the glucose units in this macromolecule are joined by beta-glycoside bonds between C-1 and C-4 sites of adjacent sugars. Partial hydrolysis of starch and glycogen produces the disaccharide maltose together with low molecular weight dextrans, polysaccharides in which glucose molecules are joined by alpha-glycoside links between C-1 and C-6, as well as the alpha C-1 to C-4 links found in maltose.

Over half of the total organic carbon in the earth's biosphere is in cellulose. Cotton fibres are essentially pure cellulose and the wood of bushes and trees is about 50% cellulose. As a polymer of glucose, cellulose has the formula (C6H10O5)n where n ranges from 500 to 5,000, depending on the source of the polymer. The glucose units in cellulose are linked in a linear fashion, as shown in the drawing below. The beta-glycoside bonds permit these chains to stretch out and this conformation is stabilized by intramolecular hydrogen bonds. A parallel orientation of adjacent chains is also favoured by intermolecular hydrogen bonds. Although an individual hydrogen bond is relatively weak, many such bonds acting together can impart great stability to certain conformations of large molecules.

Most animals cannot digest cellulose as a food and in the diets of humans this part of our vegetable intake functions as roughage and is eliminated largely unchanged. Some animals (the cow and termites, for example) harbour intestinal microorganisms that breakdown cellulose into monosacharide nutrients by the use of beta-glycosidase enzymes.

Cellulose is commonly accompanied by a lower molecular weight, branched, amorphous polymer called hemicellulose. In contrast to cellulose, hemicellulose is structurally weak and is easily hydrolyzed by dilute acid or base. Also, many enzymes catalyze its hydrolysis. Hemicelluloses are composed of many D-pentose sugars, with xylose being the major component. Mannose and mannuronic acid are often present, as well as galactose and galacturonic acid.

Starch is a polymer of glucose, found in roots, rhizomes, seeds, stems, tubers and corms of plants, as microscopic granules having characteristic shapes and sizes. Most animals, including humans, depend on these plant starches for nourishment. The structure of starch is more complex than that of cellulose. The intact granules are insoluble in cold water, but grinding or swelling them in warm water causes them to burst. The released starch consists of two fractions. About 20% is a water soluble material called amylose. Molecules of amylose are linear chains of several thousand glucose units joined by alpha C-1 to C-4 glycoside bonds.

Amylose solutions are actually dispersions of hydrated helical micelles. The majority of the starch is a much higher molecular weight substance, consisting of nearly a million glucose units and called amylopectin. Molecules of amylopectin are branched networks built from C-1 to C-4 and C-1 to C-6 glycoside links and are essentially water insoluble. The branching in this diagram is exaggerated, since on average, branches only occur every twenty five glucose units. Hydrolysis of starch, usually by enzymatic reactions, produces a syrupy liquid consisting largely of glucose. When cornstarch is the feedstock, this product is known as corn syrup. It is widely used to soften texture, add volume, prohibit crystallization and enhance the flavour of foods.

Glycogen is the glucose storage polymer used by animals. It has a structure similar to amylopectin, but is even more highly branched (about every tenth glucose unit). The degree of branching in these polysaccharides may be measured by enzymatic or chemical analysis.

SYNTHETIC MODIFICATION OF CELLULOSE

Cotton, probably the most useful natural fibre, is nearly pure cellulose. The manufacturie of textiles from cotton involves physical manipulation of the raw material by carding, combing and spinning selected fibres. For fabrics the best cotton has long fibres and short fibres or cotton dust are removed. Crude cellulose is also available from wood pulp by dissolving the lignan matrix surrounding it. These less desirable cellulose sources are widely used for making paper.

In order to expand the ways in which cellulose can be put to practical use, chemists have devised techniques for preparing solutions of cellulose derivatives that can be spun into fibres, spread into a film or cast in various solid forms. A key factor in these transformations are the three free hydroxyl groups on each glucose unit in the cellulose chain, $-[C_6H_7O(OH)_3]_n$. Esterification of these functions leads to polymeric products having very different properties compared with cellulose itself.

Cellulose Nitrate, first prepared over 150 years ago by treating cellulose with nitric acid, is the earliest synthetic polymer.

The fully nitrated compound, $-[C_6H_7O(ONO_2)_3]_n$, called guncotton, is explosively flammable and is a component of smokeless powder.

Partially nitrated cellulose is called pyroxylin. Pyroxylin is soluble in ether and at one time was used for photographic film and lacquers. The high flammability of pyroxylin caused many tragic cinema fires during its period of use. Furthermore, slow hydrolysis of pyroxylin yields nitric acid, a process that contributes to the deterioration of early motion picture films in storage.

Cellulose Acetate, $-[C_6H_7O(OAc)_3]_n$, is less flammable than pyroxylin and has replaced it in most applications. It is prepared by reaction of cellulose with acetic anhydride and an acid catalyst.

The properties of the product vary with the degree of acetylation. Some chain shortening occurs unavoidably in the preparations. An acetone solution of cellulose acetate may be forced through a spinnerette to generate filaments, called acetate rayon, that can be woven into fabrics.

Viscose Rayon, is prepared by formation of an alkali soluble xanthate derivative that can be spun into a fibre that reforms the cellulose polymer by acid quenching.

STRUCTURE OF CARBOHYDRATES

MONOSACCHARIDES STRUCTURES

Sugars can be defined as polyhydroxy aldehydes or ketones. Hence the simplest sugars contain at least three carbons. The most common are the aldo- and keto-trioses, tetroses, pentoses and hexoses. The simplest 3C sugars are glyceraldehye and dihydroxyacetone.

Glucose, an aldo-hexose, is a central sugar in metabolism. It and other 5 and 6C sugars can cyclize through intramolecular nucleophilic attack of one of the OH's on the carbonyl C of the aldehyde or ketone.

Such intramolecular reactions occur if stable 5 or 6 member rings can form. The resulting rings are labeled furanose (5 member) or pyranose (6 member) based on their similarity to furan and pyran. On nucleophilic attack to form the ring, the carbonyl O becomes an OH which points either below the ring (a anomer) or above the ring (β anomer).

Monosaccharides in solution exist as equilbrium mixtures of the straight and cyclic forms. In solution, glucose (Glc) is mostly in the pyranose form, fructose is 67% pyranose and 33% furanose and ribose is 75% furanose and 25% pyranose.

However, in polysaccharides, Glc is exclusively pyranose and fructose and ribose are furanoses. Sugars can be drawn in the straight chain form as either Fisher projections or perspective structural formulas.

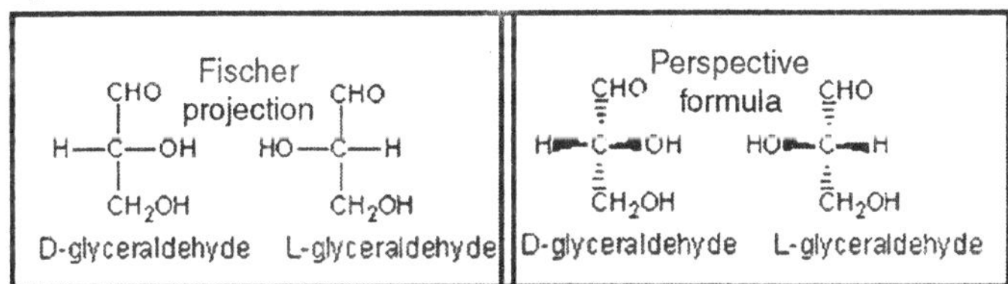

Cyclic forms can be drawn either as the Haworth projections, which shows the molecule as cyclic and planar with substituents above or below the ring) or the more plausible bent forms (showing Glc in the chair or boat conformations, for example).

b-D-glucopyranose is the only aldohexose which can be drawn with all its bulky substituents (OH and CH_2OH) in equatorial positions, which probably accounts for its widespread prevalence in nature.

Haworth projections are more realistic than the Fisher projections, but you should be able to draw both structures. In general, if a substituents points to the right in the Fisher structure, it points down in the Haworth. if it points left, it points up. In general, the OH on the a-anomer points down (ants down) while on the b-anomer it points up (butterflys up).

Fig. A more Rigorous view of the Relationship Between the Anomeric OH and the OH on the Last Chiral C of a Sugar

The most common monosaccharides (other than glyceraldehyde and dihydroxyacetone) which you need to know are shown below.

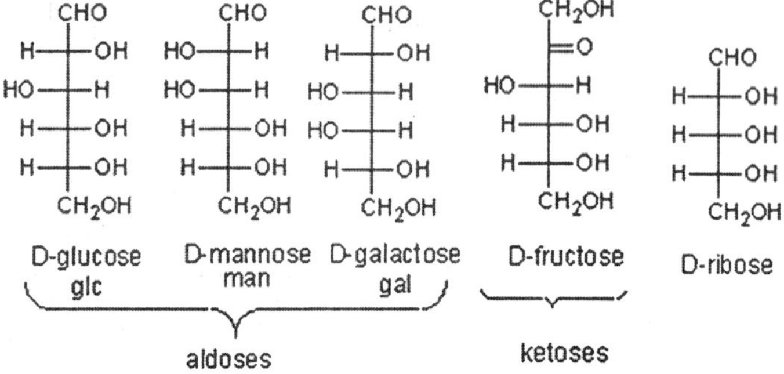

The mirror image of D-Glc is L-Glc. For common sugars, the prefix D and L refer to the cenre of asymmetry most remote from the aldehyde or ketone. By convention, all chiral centres are related to D- glyceraldehyde, so sugar isomers related to D-glyceraldehyde at their last asymmetric cenre are D sugars.

ISOMERS

Sugars can exists as either configurational isomers (interconverted only by breaking covalent bonds) and conformational isomers.

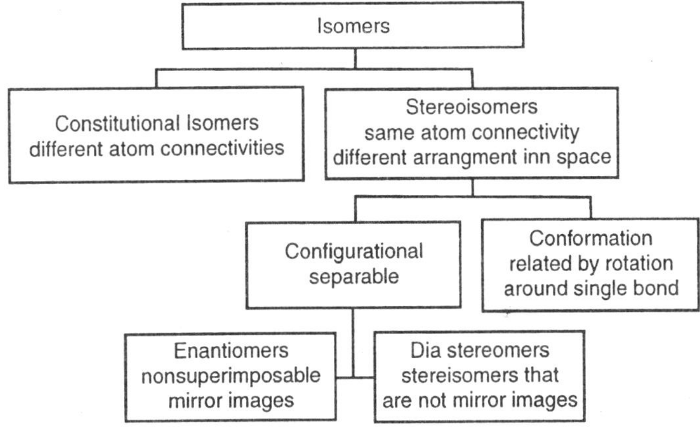

Fig. Different Types of Isomers

The configurational isomers include enantiomers (stereoisomers that are mirror images of each other), diastereomers (stereoisomers that are not mirror images), epimers (diastereomers that differ at one stereocentre) and anomers (a special form of stereoisomer, diastereomer and epimer that differ only in the configuration around the carbon which was attacked in the intramolecular nucleophilic attack to produces the a and b isomers).

Conformational Isomers

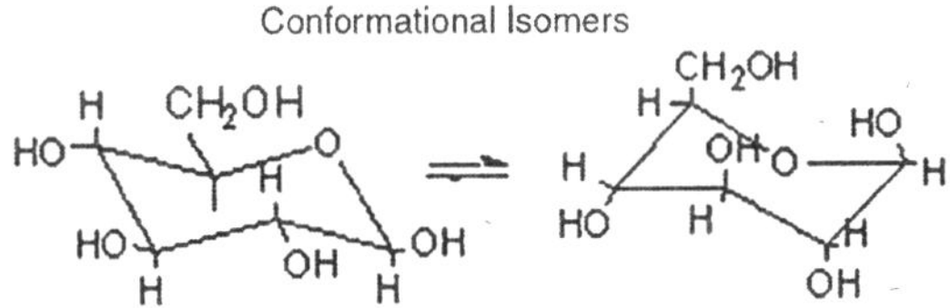

β-D-glucopyranose: chairnforms, boat form also is conformer

Monosaccharide Derivatives

Many derivatives of monosaccharides are found in nature.

These include:

- Oxidized forms in which the aldehyde and/or alcohol functional groups are oxidized to carboxylic acids
- Phosporylated forms in which phosphate is added by ATP to form phosphoester derivatives

- Amine derivatives such as glucosamine or galactosamine
- Acetylated amine derivatives such as N-Acetyl-GlcNAc (GlcNAc) or GalNAc
- Lactone forms (intramolecular esters) in which an OH group attacks a carbony C that was previously oxidized to a carboxylic acid
- Condensation products of sugar derivatives with lactate ($CH_3CHOHCO_2^-$) and pyruvate, ($CH_3COCO_2^-$), both from the glycolytic pathway, to form muramic acid and neuraminic acids, (also called sialic acids), respectively.

N-acetylmuramic acid, found in bacterial cell walls, consists of GlcNAc in ether link at C3 with lactate, while N-acetylneuraminic acid results from a intramolecular cyclization of an condensation product of ManNac and pyruvate.

SUGARS-HEMIACETALS, ACETALS, DISACCHARIDES,

Monosaccharides that contain aldehydes can cyclize through intramolecular nucleophilic attack of an OH at the carbonyl carbon in an addition reaction to form a hemiacetal (hemiketal if attack on a ketone).

Aldehydes, Hemiacetals, and Acetals

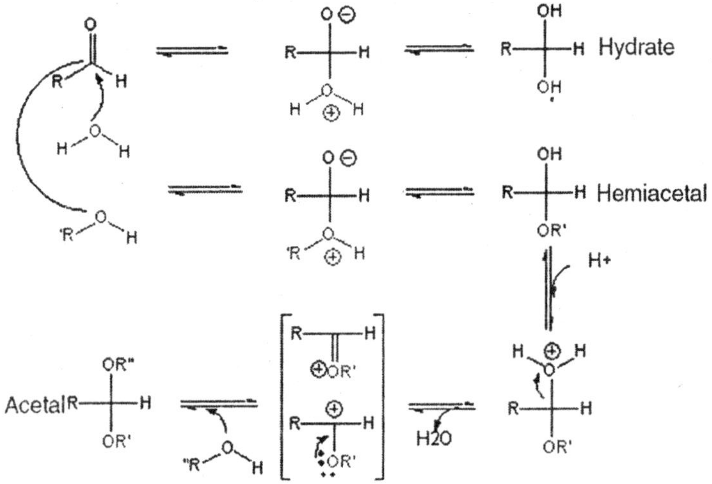

Hydrate

Hemiacetal

Acetal

On the addition of acid (which protonates the anomeric OH, forming water as a potential leaving group), another alcohol can add forming an acetal (or ketal from a ketone) with water leaving.

If the other alcohol is a second monosaccharide, a dissacharide results. The acetal (or ketal) link bonding to the two monosaccharides is called a glycosidic link.

Links between the two sugars can be either a (if the OH on C1 involved in the glycosidic link is pointing down) or b (if the O on C1 involved in the glycosidic link is pointing up).

Since sugars contain so many OH groups which can act as the "secondΔ alcohol in acetal (or ketal) formation, links between sugars can be quite diverse. These include a and b forms of 1-2, 1-3, 1-4, 1-5, 1-6, 2-2, etc. links.

For example:

- *Lactose:* Gal(β 1->4)Glc Since Glc is attached to Gal through the OH on C4, its anomeric carbon, C1, could revert to the noncyclic aldehyde form. This aldehyde is susceptible to oxidation by reagents (Benedicts Solution - with citrate, Fehlings reagent - with tartrate) which are subsequently reduced. In both reagents, reducing sugars reduce a basic blue solution of $CuSO_4$ (Cu^{2+}) to form a brick red precipate of Cu_2O (Cu^+). Sugars (monosaccharides, dissacharides and polysaccharides which can form an aldehyde at C1 or have an a-hydroxymethyl ketone group which can isomerize to an aldehyde under basic conditions (such as fructose) are called reducing sugars. These oxidizing agent are mild and react with aldehydes and not ketones.
- *Sucrose:* Glc(α 1->2)Fru. Since Fru is attached through the anomeric OH of this ketose, the Fru is not in equilibrium with its straight chain keto form and hence sucrose is a nonreducing sugar.

The glycosidic (acetal or ketal) link can be cleaved by hydrolysis, just as the peptide bond in proteins.

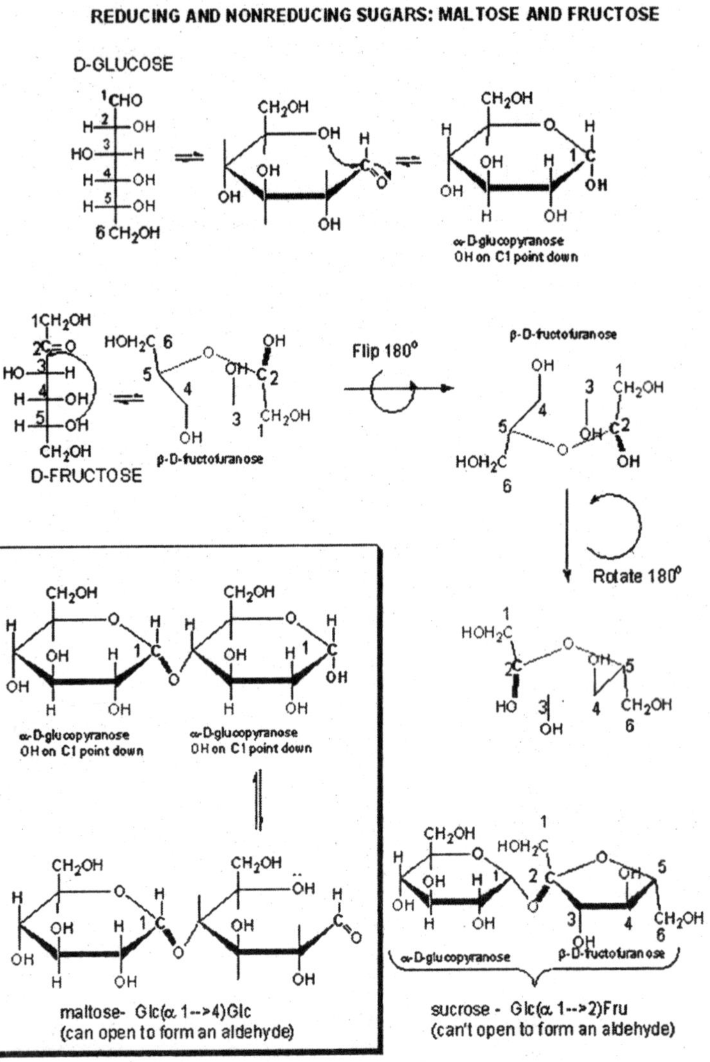

REDUCING AND NONREDUCING SUGARS: MALTOSE AND FRUCTOSE

D-GLUCOSE

α-D-glucopyranose
OH on C1 point down

D-FRUCTOSE

β-D-fuctofuranose

Flip 180°

Rotate 180°

α-D-glucopyranose
OH on C1 point down

maltose- Glc(α 1-->4)Glc
(can open to form an aldehyde)

α-D-glucopyranose β-D-fuctofuranose

sucrose - Glc(α 1-->2)Fru
(can't open to form an aldehyde)

CLASSIFICATION OF CARBOHYDRATE

Carbohydrates are classified generally according to their degree of complexity. Hence, the free sugars such as glucose and fructose are termed monosaccharides; sucrose and maltose, disaccharides; and the starches and celluloses, polysaccharides. Carbohydrates of short chain lengths such as raffinose, stachyose and verbascose, which are three, four and five sugar polymers respectively, are classified as oligosaccharides.

PENTOSES

Pentoses are five-carbon sugars seldom found in the free state in nature. In plants they occur in polymeric forms and are collectively known as pentosans. Thus, xylose and arabinose are the constituents of pentosans present in plant fibres and vegetable gums, respectively. As the sugar moieties in nucleic acids and riboflavin, ribose and deoxyribose are indispensable constituents of the life process. D-ribose has the following chemical structure.

HEXOSES

The hexoses comprise a large group of sugars. Principal among these are: glucose, fructose, galactose and mannose. While glucose and fructose are found free in nature, galactose and mannose occur only in combined form. The hexoses are divided into aldoses and ketoses according to whether they possess aldehydic or ketonic groups. Thus, glucose is an aldo sugar and fructose is a keto sugar.

The presence of aymmetric centres in all sugars with three or more carbon atoms gives rise to stereoisomers. Galactose and mannose are stereoisomers of glucose which, theoretically, is only one of 16 stereoisomers. Because the ketohexoses have only three asymmetric centres, fructose is one of eight stereoisomers.

A general phenomenon, known as mutarotation, is observed in a variety of pentoses and hexoses as well as in certain disaccharides. For example, it has been established that two isomers of D-glucose exist, hence requiring an additional asymmetric centre in this sugar. It became apparent that D-glucose and most other sugars have cyclic structures.

The position of the hydroxyl group in relation to the ring oxygen characterizes this additional configurations modification. By convention, the positioning of the hydroxyl group on carbon atom 1 on the same side of the structure as the oxygen ring -modification; and, the positioning of the same hydroxyl group on the opposite side of the ring oxygen b-modification.

Carbohydrases, which catalyse the hydrolysis of glycosidic linkages of simple glycosides, oligosaccharides and polysaccharides often exhibit specificity with regard to substrate configuration. The specificity for enzyme hydrolysis of certain oligosaccharides helps to explain the poor utilization of this class of carbohydrates in fish nutrition.

Sugars containing the aldo or the keto group are capable of reducing copper in alkaline solutions to produce the brick-red colouration of cuprous ions. These sugars are called reducing sugars and the reaction, although not specific for reducing sugars, has use for both qualitative and quantitative determinations.

Glucose is widely distributed in small amounts in fruits, plant juices and honey. It is commercially produced by the acid or enzyme hydrolysis of grain and root starches. Glucose is of special interest in nutrition because it is the end-product of carbohydrate digestion in all non-ruminant animals including fish. Fructose is the only important ketohexose and is found in the free state alongside glucose in ripening fruits and honey. Combined with glucose it forms sucrose. Fructose is somewhat sweeter than sucrose and is produced in increasing quantities commercially as a sweetener.

Galactose occurs in milk in combination with glucose. It is also present in oligo-saccharides of plant origin, in combination with both glucose and fructose.

Mannose is present in some plant polysaccharides collectively termed mannans.

DISACCHARIDES

Disaccharides are condensation products of two molecules of monosaccharides. Sucrose is the predominant disaccharide occurring in the free form and is the principal substance of sugar cane and sugar beet. It is also formed during germination of legume seeds. Other common disaccharides are maltose and lactose. Maltose is a dimer of glucose and lactose is a copolymer of galactose and glucose. The two molecules of glucose in maltose are held together in an a -1,4 glycosidic linkage whereas the two hexose entities of galactose are linked at the b -1,4 position. Glucose and fructose are combined in an a -1,2 linkage in sucrose. The abbreviated name of sucrose is D-Glu-(a, 1® 2)-D-Fru.

OLIGOSACCHARIDES

The Oligosaccharides raffinose, stachyose and verbascose are present in significant quantities in legume seeds. Raffinose, which is the most widespread among the three, consists of one molecule of glucose linked to a molecule of sucrose at the α-1, 6 position.

Its abbreviated chemical name is α-D-Gal (1→6) -α - D -Glu - (1 → 2) - β-D-Fru. Further chain elongation at the galactose end with another galactose molecule will yield stachyose.

These galactose-galactose linkages are all at the α-l,6 position and digestion of these Oligosaccharides by animals requires a highly specific enzyme not elaborated by the animals themselves but by certain bacteria present in the animals guts.

The gradual disappearance of oligosaccharides from the cotelydons of legume seeds during germination is part of an intricate process beginning with uptake of water by the seed. This uptake of moisture releases gibberellic acid which in turn activates the DNA in the seed, thereby triggering the life cycle of the plant.

The DNA directs the production of α-galactosidase which is required for the hydrolysis of these Oligosaccharides. Any interference of the DNA transcription process blocks enzyme production and will be evidenced by continued senescence of the seed and persistence of oligosaccharides in the seed cotelydons.

POLYSACCHARIDES

The polysaccharides represent a large group of complex

carbohydrates which are condensation products of undetermined numbers of sugar molecules. The various subgroups are rather ill-defined and there is a lack of agreement on their classification. Most polysaccharides are insoluble in water. Upon hydrolysis with acids or enzymes they eventually yield their constituent monosaccharides.

Starch is a high molecular weight polymer of D-glucose and is the principal reserve carbohydrate in plants. Most starches consist of a mixture of two types of polymers, namely; amylose and amylopectin.

The proportion of amylose and amylopectin is generally one part of amylose and three parts of amylopectin. Enzymes capable of catalyzing the hydrolysis of starch are present in the digestive secretions of animals and fish within their cells.

The a-amylases which are found virtually in all living cells cleave the α-D-(1→4) linkages at random and bring about an eventual total conversion of the starch molecule into the reducing sugars. The principal α-amylases of animal origin are those produced in the salivary gland and the pancreas. Starch is insoluble in water and is stained blue by iodine.

Glycogen is the only complex carbohydrate of animal origin. It exists in limited quantities in liver and muscle tissues and acts as a readily available energy source. Dextrins are intermediate compounds resulting from incomplete hydrolysis or digestion of starch.

The presence of α-D-(1→6) linkages in amylopectin and the inability of α-amylase to cleave these bonds give rise to low molecular weight carbohydrate segments called limit dextrins. These residues are acted upon primarily by acidophilic bacteria in the digestive tract.

Cellulose is made up of long chains of glucose units held together by β-D-(1→4) linkages. The enzymes which cleave these linkages are not ordinarily present in the digestive secretions of animals and fish although some species of shellfish are believed to elabourate cellulase, the enzyme which catalyzes the hydrolysis of cellulose.

Cellulase producing micro-organisms present in the gut of herbivorous animals and fish impart to their host animals the ability to utilize as food the otherwise indigestible cellulose.

Other complex polysaccharides in common occurrence are the hemicelluloses and pentosans. Hemicellulose represents a group of carbohydrates including araban, xylan, certain hexosans and polyuronides.

These substances are generally less resistant to chemical treatment and undergo some degree of enzymatic hydrolysis during normal digestive processes. Pentosans are polymers of either xylose or arabinose as constituents of plant structural material and vegetable gums, respectively.

Arbohydrates are classified generally according to their degree of complexity. Hence, the free sugars such as glucose and fructose are termed monosaccharides; sucrose and maltose, disaccharides; and the starches and celluloses, polysaccharides.

Carbohydrates of short chain lengths such as raffinose, stachyose and verbascose, which are three, four and five sugar polymers respectively, are classified as oligosaccharides.

GLYCOPROTEINS

For many years glycoproteins have been a subject of interest. However, it is in the second half of this century that they have aroused the interest of biochemists and biologists from a wide range of fields.

This increased interest is partly due to the fact that glycoproteins were discovered to be abundant in living organisms.

It is also due to the diverse functions of glycoproteins; glycoproteins appear in nearly every biological process studied. Many glycoproteins have structural functions. One of many instances is their role as a constituent of the cell wall.

Glycoproteins also form connective tissues such as collagen. They are also found in gastrointestinal mucus secretions. Glycoproteins are used as protective agents and lubricants. They are also found abundantly in the blood plasma where they serve many functions.

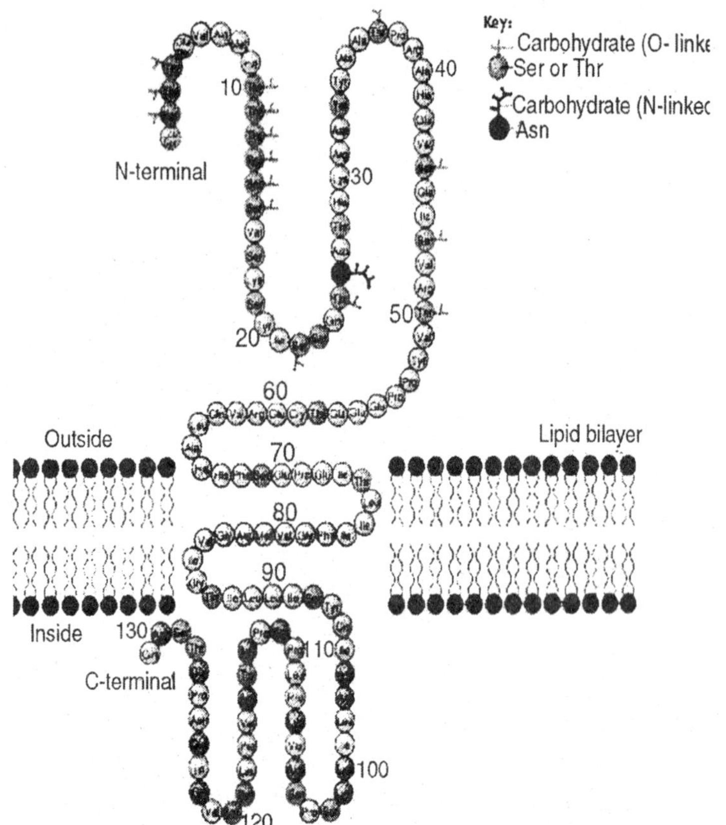

Fig. Structure of Glycoproteins

The diverse function of glycoproteins is a direct result of their structure. These macromolecules are composed of a peptide chain with one or more carbohydrate moieties. There are two broad categories of glycoprotein structure. The carbohydrates are either linked N-glycosidically or O-glycosidically to their constituent protein. Within these broader categories, there can be fine structural differences which account for the large diversity of functions among glycoproteins. Controlling of glycoproteins is achieved through synthesis and degradation. Those processes are controlled by very specific enzymes. Not so much is yet researched on glycoprotein regulation in general.

STRUCTURE

Structurally, glycoproteins consist of a polypeptide covalently bonded to a carbohydrate moiety. The carbohydrate can make up anywhere from less than one per cent to more than 80 per cent of the total protein mass. The saccharide chains, referred to as glycans, can be linked to the polypeptide in two major ways.

The first class of glycoproteins are the O-linked glycans. These usually contain an N-acetylgalactosamine which is attached through a glycosidic bond to the O-terminus of either threonine or serine. The other class of glycoproteins are the N-linked glycans. These involve a glycosidic bond between N-acetylglucosamine and the N-terminus of an asparagine residue.

O-linked glycans consist of N-acetylgalactosamine attached to the O-terminus of a threonine or serine residue. N-acetylgalactosamine is simply a galactose molecule with an amine group covalently bonded to the second carbon. This amine group is bonded to a carboxyl group. N-acetylgalactosamine attaches to the carboxyl group of the amino acid through the hydroxyl group of its anomeric carbon.

Another type of O-linked glycan consists of a galactose or a glucosyl-galactose disaccharide linked to the hydroxyl of hydroxylysine. Yet another type of O-linkage involves the binding of arabinose to the hydroxyl of hydroxyproline. In all of the 0-linked glycans, there can be a variety of different monosaccharide or polysaccharide chains attached to the sugar that is bonded to the amino acid.

The other class of glycoproteins are N-linked glycans. These molecules consist of an N-acetylglucosamine bonded to the amide nitrogen of an asparagine molecule. N-acetylglucosamine is simply a glucose molecule which is bonded to an amine group.

This amine group, in turn, is bonded to a hydroxyl group. The N-acetylglucosamine is bonded to the asparagine through its anomeric carbon. The asparagine must be surrounded by a specific amino acid sequence, or sequon.

This sequence is -X-Asn-X-Thr; the X can be any amino acid. A large variety of polysaccharide side chains can be linked to the N-acetylglucosamine. A typical polysaccharide chain is Man2 a(1-6)-Man B(1-4)-GlcNAc B(1-4)-GlcNAc B(1-N) Asn. Adding on to this structure can create many different N-linked glycans. The carbohydrate chains of glycoproteins can play a role in the structure of the polypeptide. For example, in human immunoglobulins, the carbohydrate chain wraps around one of the protein domains.

By doing so it prevents contact of this domain with the neighbouring domain. An experiment that was done by Koide et al illustrates how the carbohydrate can change the overall structure of the polypeptide. The carbohydrate side chains of a rabbit antibody were removed through glycosidase digestion.

The result was that the domain where the carbohydrate had previously been attached could no longer perform its ordinary function. Because an immunoglobulin's function is determined to a large extent by its structure, it can be concluded that removing the carbohydrate affected the structure of the molecule.

FUNCTION

Because carbohydrates and proteins by themselves serve in a vast number of biological functions, it should not be surprising that linking the two together results in a macromolecule with an extremely large number of functions. Because of this and their biologically ubiquitous nature, the best way to go about exploring glycoprotein function is to break it down into categories that are fairly general. The following is an attempt to do this.

Structural: Glycoproteins are found throughout matrices. They act as receptors on cell surfaces that bring other cells and proteins together giving strength and support to a matrix

Proteoglycan: Linking glycoproteins cross links proteoglycan molecules and is involved in the formation of the ordered structure within cartilage tissue. In nerve tissue glycoproteins are abundant in gray matter and appear to be associated with synaptosomes, axons and microsomes. Prothrombin, thrombin and fibrinogen are all glycoproteins that play an intricate role in the blood clotting mechanism . In certain bacteria the slime layer that surrounds the outermost components of cell walls are made up of glycoproteins of high molecular weight.

In addition to forming these s-layers, glycoproteins also function as bacterial flagella. These are made up of bundles of glycoproteins protruding from the cell's surface. Their rotation provides propulsion. In plants, glycoproteins have roles in cell wall formation, tissue differentiation, embryogenesis and sexual adhesion

Protection: High molecular weight polymers called mucins are found on internal epithelial surfaces. They form a highly viscous gel that protects epithelium form chemical, physical and microbial disturbances. Examples of mucin sites are the human digestive tract, urinary tract and respiratory tracts. "Cervical mucinΔ is a glycoprotein found in the cervix of animals that regulates access of spermatozoa to the upper reproductive tract.

Recently it was discovered that mucins may be responsible for aiding in metastasis of transformed cancer cells Mucins are also found on the outer body surfaces of fish to protect the skin Not only does mucin serve the function of protection, but it also acts as a lubricant. Human lacrimal glands produce a glycoprotein which protects the corneal epithelium from desiccation and foreign particles. Human sweat glands secrete glycoproteins which protect the skin from the other excretory products that could harm the skin.

Reproduction: Glycoproteins found on the surface of spermatozoa appear to increase a sperm cell's attraction for the egg by altering the electrophoretic mobility of the plasma membrane.

Actual binding of the sperm cell to the egg is mediated by) -linked glycoproteins serving as receptors on the surface of each the two membranes. The zona pellucida is an envelope ′made of glycoprotein that surrounds the egg and prevents polyspermy from occurring after the first sperm cell has penetrated the egg's plasma membrane Hen ovalbumin is a glycoprotein found in egg white that serves as a food storage unit for the embryo

Adhesion: Glycoproteins serve to adhere cells to cells and cells to substratum. Cell-cell adhesion is the basis for the development of functional tissues in the body. The interactions between cells is mediated by the glycoproteins on those cell's surfaces.

In different domains of the body, different glycoproteins act to unite cells. For example, nerve cells recognize and bind to one another via the glycoprotein N-CAM (nerve cell adhesion molecule). N-CAM is also found on muscle cells indicating a role in the formation of myoneural junctions.

With cell-substratum adhesion, glycoproteins serve as cell surface receptors for certain adhesion ligands that mediate and coordinate the interaction of cells. Substrates with the appropriate receptor will bind to the cell related to that receptor.

For example, a substrate containing the glycoprotein fibronectin will be recognized and adhered to by fibroblasts. The fibroblasts will then secrete adhesion molecules and continue to spread, producing a pericellular matrix .

Hormones: There are many glycoproteins that function as hormones such as human chorionic gonadotropin (HCG) which is present in human pregnancy urine. Another example is erythropoietin which regulates erythrocyte production

Enzymes: Glycoprotein enzymes are of three types. These are oxidoreductases, transferases and hydrolases

Carriers: Glycoproteins can bind to certain molecules and serve as vehicles of transport. They can bind to vitamins, hormones, cations and other substances.

Inhibitors: Many glycoproteins in blood plasma have shown antiproteolytic activity. For example, the glycoprotein a1-antichymotrypsin inhibits chymotrypsin.

Defence: In beetles pygidial glands secrete a glycoprotein disinfecting paste that covers the body and hardens. This shell provides protection against attack by bacteria and fungi

Freezing-point depression: Glycoproteins were found in the sera of antarctic fishes to decrease the freezing point due to their apparent interaction with water

Vision: In bovine visual pigment a glycoprotein forms the outer membranes of retinal rods

Immunological: The interaction of blood group substances with antibodies is determined by the glycoproteins on erythrocytes. Adding or removing just one monosaccharide from a blood group structure, the antigenicity and therefore a person's blood type can be altered Many immunoglobulins are actually glycoproteins Soluble immune mediators such as helper, suppressor and activator cell have been shown to bind to glycoproteins found on the surface of their target cells. B and T cells contain surface glycoproteins that attract bacteria to these sites and bind them. In much the same manner, glycoproteins can direct phagocytosis. Because the HIV virus recognizes the receptor protein CD4, it binds to helper T cells which contain it.

REGULATION/CONTROL

Regulation and control of glycoproteins is not as straight forward as some might think. "If someone was to understand regulation in glycoproteins he would have to look at the enzymes that are involved in the biosynthesis pathway of these molecules; what is affecting the enzyme activity (could be hormonal). The big interest in this area is in the cloning of the genes, the cloning of the enzymes that are involved in the synthesis of the oligosaccharide on the glycoproteins.

SYNTHESIS

The protein part of the glycoprotein is formed at the ribosomes, where all proteins are synthesized, on template represented by RNA and DNA. As a result, its structure can change only through the mutation of the genetic material of the cell. On the other hand, the carbohydrate component of a glycoprotein is not a product of the ribosome; it is synthesized somewhere in the cytoplasm, but the exact site of synthesis has not yet been established. Since it is not directly genetically controlled, the oligosaccharide part shows a much greater variation

In contrast with O-linked glycoproteins, where oligosaccharide assemble occurs on the polypeptide chain, N-linked glycoproteins assemble their oligosaccharide portions on a lipid linked intermediate, dolichol phosphate. The first step in oligosaccharide synthesis is the formation of that intermediate. In subsequent steps, sugars, the first being always N-acetylglucosamine are chain-like connected to dolichol phosphate. One more GlcNAc and three more mannose sugars are linked to the first GlcNAc to form the typical core of N-linked glycoproteins. Addition of any other sugar can be in any possible combination, according to the desired resulting functions. The specificity of the enzymes is very important in the synthesis process. Every sugar added, is catalyzed by a different enzyme.

These group of enzymes are called glycosyltransferases The first addition of GlcNAc to dolichol phosphate is catalyzed by the specific enzyme example) that cleave peptide bonds and glycosidases enzymes that remove sugars one at a time from the end of an oligosaccharide chain. Both of these groups are contained in lysozomes.

The lysosome attaches to a phagocyte, which has engulfed a substance that needs to be broken down and releases its enzymes in it Next, these enzymes begin their catalytic action. In degradation, ad in synthesis, the enzymes involved are very specific. After the glycoprotein is broken down, its amino acid and sugar components are either metabolized or can be used in the formation of another glycoprotein.

Enzymatic degradation can provide much information about the structure of the oligosaccharide chains, as well as about the carbohydrate peptide linkage. For example, if a glycoprotein is treated with mannosidase (removes mannose) and mannose is released, one can conclude that mannose residues were located at the periphery of the molecule since glycosidases remove sugars from the end of the oligosaccharide chain

Regulation and control is the organism's ultimate tool to monitor and adjust the production or degradation of different molecules. Like a factory, it would be useless and inefficient to produce excess products when there is no need for them. On the other hand, shortage of products could be a problem too. Hopefully, further research will put some light in the area of glycoprotein regulation.

GLYCOPROTEIN

Glycoproteins play an important part in hormone function. The action of hormones depends on the initial binding of the hormone to a protein receptor molecule. In many cases this molecule is a glycoprotein. Many hormones bind to receptors in the cell membrane; these hormones never actually enter the cell.

Steroids, on the other hand, bind to an intracellular protein receptor. There is still controversy over whether the steroid hormone receptor is found in the nucleus or the cytosol. However, it is clear that after the steroid binds, the hormone-receptor complex moves to the nucleus.

The hormone binding domain of the receptor is found at the C-terminus. The amino acid sequence of this region is highly diverse; it is not conserved from one protein to the next. Binding of the hormone stimulates a conformational change in the hormone receptor. This change allows the hormone-receptor complex to bind to the DNA.

The DNA binding domain is highly conserved and is found within the central core of the protein. The central core contains a very basic amino acid sequence. In the cellular environment, these bases will tend to pick up a hydrogen and become positively charged. The positive charge will attract the hormone-receptor complex to the negatively charged DNA. Binding of the complex to the DNA stimulates transcription

GLYCOPROTEINS, CYTOCHROMES AND METALLOENZYMES

Cytochromes have features that are quite similar to

transferrins, a particular class of glycoproteins. First of all, cytochromes make up some of the integral proteins found in membranes, as do glycoproteins. Second of all, the main function of cytochromes is the binding of ion cations. Via oxidoreduction with these cations of 2+state, cytochromes bind iron to a porphyrin prosthetic group, forming a heme group The function is to transport cations in and out of the cell. Transferrin is an iron binding protein as well.

It interacts with iron cations of the 2+state which also bind to a porphyrin group, Protoporphyrin IX, to form a heme group. Transferrins carry ions to many different locations in the body including in and out of cells Another group of molecules that serve a function related to glycoproteins are the metalloenzymes. One class of glycoprotein enzymes, the oxidoreductases, contain an enzyme called chloroperoxidase. Like transferrin, chloroperoxidase contains protophyrin IX and forms the heme group with iron 3+ Eicosanoids, are also related to glycoproteins because they recognize the receptors on cells which are glycoproteins.

ASPARAGINE LINKED GLYCOPROTEINS

Synthesis of N-linked glycoproteins can be inhibited with the use of antibiotics (function of antibiotics). Examples are monensin, mevinolin and tunicamycin. Monensin, a monocovalent ionophore, inhibits the transport of proteins through the endoplasmic reticulum- Golgi complex . Mevinolin on the other hand, is an inhibitor of mevalonate synthesis. Mevalonate is the precursor of dolichol and other isoprenoid lipids Asparagine linked glycosylation involves the first step of addition of GlcNAc to dolichol diphosphate.

This reaction is catalyzed by the enzyme UDP-GlcNAc:dolichol phosphate N-acetylglucosamine-1-phosphate transferase or GPT. Tunicamycin is a specific inhibitor of GPT, therefore it can inhibit the synthesis of all N-linked glycoproteins.

To give just an example of the effectiveness of tunicamycin, when it was injected to rats' hearts, it killed 12,

24, 50 and 60 h after the injection . Another connection could be seen with the function of the ribosomes, since the polypeptide chain of all glycoproteins.

PEPTIDOGLYCAN

Peptidoglycan, also known as murein, is a polymer consisting of sugars and amino acids that forms a mesh-like layer outside the plasma membrane of eubacteria. The sugar component consists of alternating residues of β-(1,4) linked N-acetylglucosamine and N-acetylmuramic acid residues. Attached to the N-acetylmuramic acid is a peptide chain of three to five amino acids. The peptide chain can be cross-linked to the peptide chain of another strand forming the 3D mesh-like layer. Some Archaea have a similar layer of pseudopeptidog-lycan. Peptidoglycan serves a structural role in the bacterial cell wall, giving structural strength, as well as counteracting the osmotic pressure of the cytoplasm.

A common misconcep-tion is that peptidoglycan gives the cell its shape; however, whereas peptidoglycan helps maintain the structure of the cell, it is actually the MreB protein that facilitates cell shape. Peptidoglycan is also involved in binary fission during bacterial cell reproduction. The peptidoglycan layer is substantially thicker in Gram-positive bacteria (20 to 80 nm) than in Gram-negative bacteria (7 to 8 nm), with the attachment of the S-layer. Peptidoglycan forms around 90% of the dry weight of Gram-positive bacteria but only 10% of Gram-negative strains. In Gram-positive strains, it is important in attachment roles and sterotyping purposes. For both Gram-positive and Gram-negative bacteria, particles of approximately 2 nm can pass through the peptidoglycan.

ANTIBIOTIC INHIBITION

Some antibacterial drugs such as penicillin interfere with the production of peptidoglycan by binding to bacterial enzymes known as penicillin-binding proteins or transpeptidases. Penicillin-binding proteins form the bonds between oligopeptide crosslinks in peptidoglycan.

For a bacterial cell to reproduce through binary fission, more than a million peptidoglycan subunits (NAM-NAG+oligopeptide) must be attached to existing subunits. Mutations in transpeptidases that lead to reduced interactions with an antibiotic are a significant source of emerging antibiotic resistance.

Considered the human body's *own antibiotic*, lysozymes found in tears work by breaking the â-(1,4)-glycosidic bonds in peptidoglycan and thereby destroying many bacterial cells. Antibiotics such as penicillin commonly target bacterial cell wall formation (of which peptidoglycan is an important component) because animal cells do not have cell walls.

OPOLYSACCHARIDES

LIPOPOLYSACCHARIDES

These complex compounds (LPS) are the endotoxic O-antigens found in the cell walls of Gram-negative bacteria (S-lipopolysaccharides) but are also present in one fungus, they may cause several pathophysiological symptoms, such as fever, diarrhea, blood pressure decrease, septic shock and death. A lipid part (Lipid A) forms a complex with a core polysaccharide part through a glycosidic linkage. The core part is linked to a third external region of a highly immunogenic and variable O-chain polysaccharide or O-antigen made up of repeating oligosaccharide units. The latter region of the LPS molecule is responsible for bacterial serological strain specificity.

Lipid A consists of a backbone of b-1,6-glucosaminyl-glucosamine with two phosphoester groups in the 1-position of glucosamine I and in the 4-position of glucosamine II. The 3-position of glucosamine II forms the acid-labile glycosidic linkage to the long-chain polysaccharide. The other groups are substituted with hydroxylated fatty acids as hydroxymyristate (two ester-linked and two amide-linked) and normal fatty acids. The fatty acid profiles of lipopolysaccharides differ markedly from those of the phospholipids from the same species.

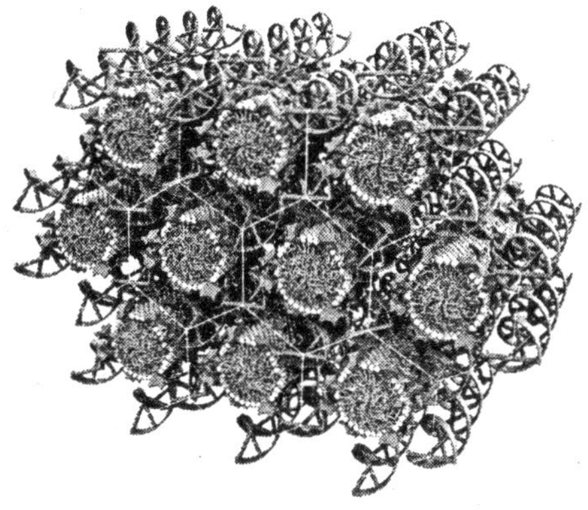

Fig. Structure of Lipid A of *Escherichia*

Although the hydroxylated fatty acids are generally 3-hydroxyalkanoic acids, 2-hydroxyalkanoic acids are also common. Exceptionally, some bacteria seem to lack hydroxy acids such as some species of *Gardnerella, Thermus, Thiobacillus, Bacteroides, Sorangium, Achromobacter* and *Brucella*. In most organisms, the major hydroxy acids are straight-chain, even-carbon compounds (from 12 to 18 carbon atoms). Odd-carbon acids are minor components (13 to 17 carbon atoms) in some enterobacteria, some *Vibrio*, some Pseudomonads and *Bordetella*. They are also found in *Pectinatus, Veillonella* and *Selenomonas*. Branched-chain hydroxy fatty acids are also present in *Pseudomonas, Moraxella, Desulfovibrio, Capnocytophaga, Flavobacterium, Legionella* species (3-OH i-11:0 up to3-OH i-17:0).

Lipopolysaccharides are the major toxins of Gram-negative bacteria (endotoxin). The only Gram-positive bacteria that possesses LPS is *Listeria monocytogenes*, an infective agent in milk. While the polysaccharide is non-toxic, the lipid A part is responsible for the toxic activities of these bacteria which result in septic shock.

CARBOHYDRATE METABOLISM IN FISH

Much of the carbohydrates that enter the diets of animals, including fish, is of plant origin. Carnivorous fish like the Atlantic salmon and the Japanese yellowtail, therefore, deal with little carbohydrate. Indeed, experiments have shown that these species are ill-equipped to handle significant quantities of raw carbohydrate, in their diets.

On the other hand, omnivores such as the common carp and the channel catfish are able to digest fair amounts of carbohydrates in their diets. The grass carp, a herbivore, subsists primarily on a vegetarian diet.

DIGESTION, ABSORPTION AND STORAGE

The ability of animals to assimilate starch depends on their ability to elabourate amylase. All species of fish have been shown to secrete a -amylase. It has also been demonstrated that activity of this enzyme was greatest in herbivores. In carnivores such as the rainbow trout and sea perch, amylase is primarily of pancreatic origin whereas in herbivores the enzyme is widespread throughout the entire digestive tract. In Tilapia mossambica the pancreas has been shown to be the site of greatest amylase activity followed by the upper intestine. Although the digestion of starch and dextrin by the carnivorous rainbow trout was shown to decrease progressively as levels of the carbohydrates were increased beyond the 20 per cent level, the fish could effectively utilize up to 60 per cent glucose, sucrose or lactose in the diet. This demonstrates that, contrary to earlier belief, carnivorous fish are capable of efficiently utilizing simple carbohydrate as a primary energy source.

The crystalline structure of starch appears also to influence its attack by amylase as evidenced by the two-fold increase in metabolizable energy content of fully cooked (gelatinized) maize in feeding trials with channel catfish. Rainbow trout have also been shown to have a higher tolerance for carbohydrate (present as wheat starch) in the diet when it was cooked. The process of gelatinization involves both heat and water. If an aqueous suspension of starch is heated, the granules do not change in appearance until a certain critical temperature is reached. At this point some of the starch granules swell and simultaneously lose their crystallinity. The critical temperature is that at which hydrogen bonds of the starch molecule loosen to permit complete hydration, leading to a phenomenon known as "swellingΔ.

Alpha-amylase, promotes a more or less random fragmentation of the starch molecule by hydrolyzing at the α -D-(l→4) glucosidic bonds in the inner and outer chains of the compound. The result of complete hydrolysis of the amylose component are maltose and D-glucose, while the amylopectin component is reduced to maltose, D-glucose and branched limit dextrins. As a consequence of these action patterns by a -amylase on starch, other enzymes are needed for complete hydrolysis of starch to D-glucose in fish. In this regard, it has been demonstrated that even the carnivorous sea bream possess the ability to digest maltose.

On the other hand, cellulase and a -galactosidase have not been shown to be secreted by fish although cellulase of bacterial origin is present in the gut of most species of carps. The lack of a -galactosidase may partly explain the poor response by fish to dietary soybean meal which contains significant levels of the galactosidic oligosaccharides raffinose and stachyose. As has been pointed out earlier, these oligosaccharides do undergo enzymatic hydrolysis during the germination process to yield galactose and sucrose.

This can be achieved by soaking the beans for 48 hours prior to processing for meal production. It should also be pointed out that the nutritive value of pulses and other legume seeds can likewise be improved for fish since oligosaccharides constitute a large portion of the carbohydrates in legume seeds. Data on glucose absorption by fish are scanty. Work with goldfish has shown that active transport of glucose is coupled with Na^+ transport as in most mammals. It is generally believed that absorption takes place on the mucosal surface of intestinal cells.

The mono-saccharides which result from carbohydrate digestion consist primarily of glucose, fructose, galactose, mannose, xylose and arabinose. Although the rates of absorption of these sugars have been determined for many land mammals, similar information for fish is not available.

FACTORS AFFECTING METABOLISM

Apart from genetic adaptation, climatic factors also play an important role in carbohydrate metabolism in fish. Acclimation in fish, in essence, reflects enzyme acclimation, since the animal's ability to survive depends largely upon its ability to carry out normal metabolic functions. Some enzymes for metabolic acclimation show good compensation while others do not.

The enzymes associated with energy liberation (enzymes of glycolysis, pentose shunt, tricarboxylic acid cycle, electron transport and fatty acid oxidation) exhibit temperature compensation whereas, those enzymes dealing largely with the degradation of metabolic products show poor or reverse compensation

Table. Enzymes Subject to Metabolic Acclimation

Enzymes exhibiting compensation	Enzymes exhibiting reverse or no compensation
Phosphofructokinase	Catalase
Aldolase	Peroxidase
Lactic dehydrogenase	Acid phosphatase
6-phosphogluconate dehydrogenase	D-amino acid oxidase
Succinic dehydrogenase	Mg-ATP ase
Malic dehydrogenase	Cholineacetyl transferase
Cytochrome oxidase	Acetylcholine esterase
Succinate-cytochrome C reductase	Alkaline phosphatase
NAD-cytochrome C reductase	Allantoinase
Aminoacyl transferase	Uricase
Na-K-ATPase	Amylase
protease	Lipase
	Malic enzyme
	glucose-6-phosphate dehydrogenase

It is interesting to note that two key enzymes involved in carbohydrate metabolism, amylase and glucose-6-phosphate dehydrogenase, together with an enzyme involved in fat digestion, lipase, show no temperature compensation. It is not certain if this is in any way connected with the cessation of feeding by fish at low temperatures. The molecular mechanism of thermal acclimation are not well understood and may consist of changes in synthesis or amounts of a given enzyme. Differences in kinetics, changes in the proportion of isoenzymes suitable for particular temperatures and changes in co-factors such as lipids, co-enzymes, or other factors such as pH and ions may be important in the animal's adjustment to temperature changes.

ENERGY TRANSFORMATION

Despite species differences in the tolerance of dietary carbohydrates it is generally believed that the principal end-product of carbohydrate digestion, glucose, is metabolized in a manner prevailing in all cells, i.e., via the reversible Emden-Meyerhoff pathway.

In this pathway, glucose has only one principal fate: phosphorylation to glucose-6-phosphate. The major metabolic transformations are depicted as follows: All transformations proceed with a loss of free energy. Thus, the formation of two moles of lactate from glucose-6-phosphate occurs with free energy change of D G^{o} = -22000 cal/mole. The net result is the formation of four molecules of ATP. A functional reversal of this transformation can only occur via a different sequence requiring the input of six ATP molecules per mole of glucose-6-phosphate recovered.

Cells do not store glucose or glucose-6-phosphate. The readily available storage form is glycogen which is made from glucose-1-phosphate by one pathway and returned by another. Although in mammalian cells glucose-6-phosphate is transformed into fatty acids, such transformation does not appear to take place in fish.

Studies with the common carp indicate that the precursor for lipogenesis is citrate formed when amino acids are actively metabolized through the tricarboxylic acid cycle. The major form of utilizable energy in all cells is ATP. In most cells this energy currency is generated by the oxidation of NADH by the mitochondrial electron-transport systems. The reductants of NAD^+ for this process are intermediates derived from the TCA cycle and fatty acids.

TCA Cycle

The TCA cycle showing enzymes, substrates and products. The abbreviated enzymes are: IDH = *isocitrate dehydrogenase* and α-KGDH = *α-ketoglutarate dehydrogenase*. The GTP generated during the *succinate thiokinase* (succinyl-CoA synthetase) reaction is equivalent to a mole of ATP by virtue of the presence of *nucleoside diphosphokinase*. The 3 moles of NADH and 1 mole of $FADH_2$ generated during each round of the cycle feed into the oxidative phosphorylation pathway.

CITRATE SYNTHASE (CONDENSING ENZYME)

The first reaction of the cycle is condensation of the methyl carbon of acetyl-CoA with the keto carbon (C-2) of oxaloacetate (OAA) to form citrate The standard free energy of the reaction, -8.0 kcal/mol, drives it strongly in the forward direction. Since the formation of OAA from its precursor is thermodynamically unfavourable, the highly exergonic nature of the *citrate synthase* reaction is of central importance in keeping the entire cycle going in the forward direction, since it *drives* oxaloacetate formation by mass action principals.

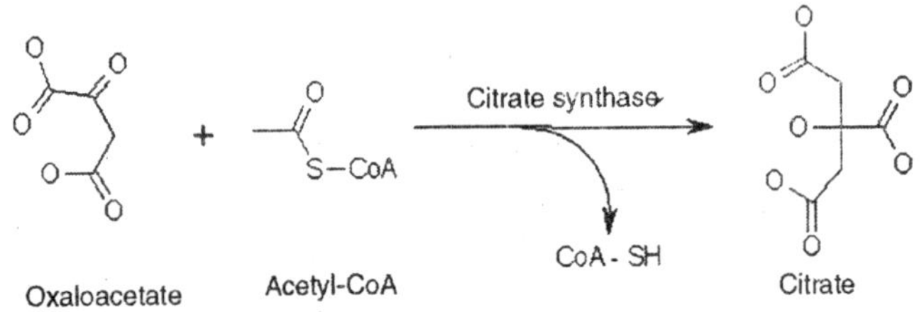

Oxaloacetate + Acetyl-CoA —(Citrate synthase)→ Citrate + CoA-SH

When the cellular energy charge increases the rate of flux through the TCA cycle will decline leading to a build-up of citrate. Excess citrate is used to transport acetyl-CoA carbons from the mitochondrion to the cytoplasm where they can be used for fatty acid and cholesterol biosynthesis. Additionally, the increased levels of citrate in the cytoplasm activate the key regulatory enzyme of fatty acid biosynthesis, acetyl-CoA carboxylase (ACC) and inhibit PFK-1. In non-hepatic tissues citrate is also required for ketone body synthesis.

ACONITASE

The isomerization of citrate to isocitrate by ***aconitase*** is stereospecific, with the migration of the -OH from the central carbon of citrate (formerly the keto carbon of OAA) being always to the adjacent carbon which is derived from the methylene (-CH_2-) of OAA.During the reaction the intermediate linked to the enzyme *cis*-aconitate is produced. The stereospecific nature of the isomerization determines that the CO_2 lost, as isocitrate is oxidized to succinyl-CoA, is derived from the oxaloacetate used in citrate synthesis.

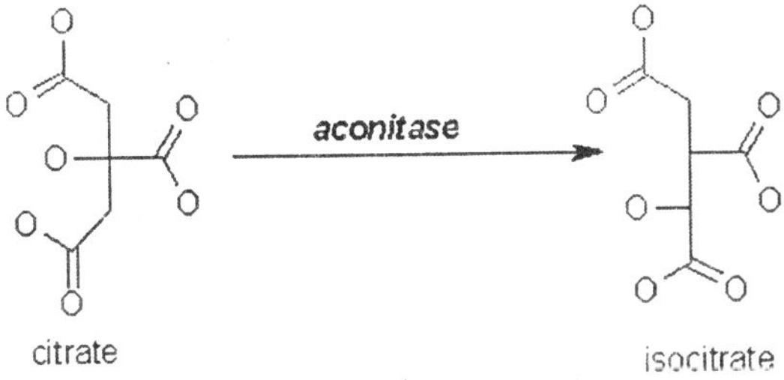

aconitase

citrate → isocitrate

Aconitase is one of several mitochondrial enzymes known as non-heme-iron proteins. These proteins contain inorganic iron and sulfur, known as iron sulfur centres, in a coordination complex with cysteine sulfurs of the protein.

There are two prominent classes of non-heme-iron complexes, those containing two equivalents each of inorganic iron and sulfur Fe_2S_2 and those containing 4 equivalents of each Fe_4S_4. Aconitase is a member of the Fe_4S_4 class. Its iron sulfur centres are often designated as $Fe_4S_4Cys_4$, indicating that 4 cystine sulfur atoms are involved in the complete structure of the complex. In iron sulfur compounds the iron is generally involved in oxidation-reduction events.

ISOCITRATE DEHYDROGENASE

Isocitrate is oxidatively decarboxylated to a-ketoglutarate by isocitrate dehydrogenase, (IDH).

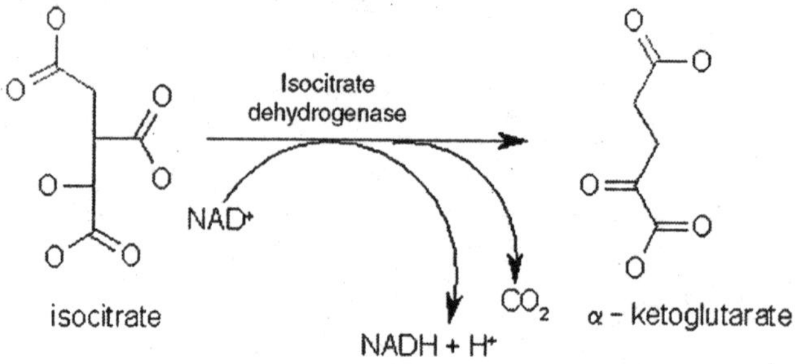

Isocitrate dehydrogenase

NAD$^+$

NADH + H$^+$

CO_2

isocitrate

α - ketoglutarate

There are two different IDH enzymes. The IDH of the TCA cycle uses NAD$^+$ as a cofactor, whereas the other IDH uses NADP$^+$ as a cofactor. Unlike the NAD+-requiring enzyme, which is located only in the mitochondrial matrix, the NADP$^+$-requiring enzyme is found in both the mitochondrial matrix and the cytosol.

IDH catalyzes the rate-limiting step, as well as the first NADH-yielding reaction of the TCA cycle. The CO_2 produced by the IDH reaction is the original C-1 of the oxaloacetate used in the citrate synthase reaction.

It is generally considered that control of carbon flow through the cycle is regulated at IDH by the powerful negative allosteric effectors NADH and ATP and by the potent positive effectors; isocitrate, ADP and AMP. From the latter it is clear that cell energy charge is a key factor in regulating carbon flow through the TCA cycle.

α-KETOGLUTARATE DEHYDROGENASE COMPLEX

α-ketoglutarate is oxidatively decarboxylated to succinyl-CoA by the *a-ketoglutarate dehydrogenase* (α-KGDH) complex.

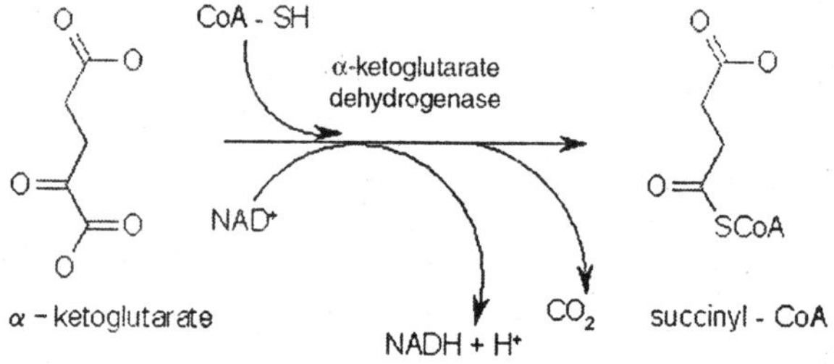

CoA - SH

α-ketoglutarate dehydrogenase

NAD$^+$

NADH + H$^+$

CO_2

SCoA

α - ketoglutarate

succinyl - CoA

This reaction generates the second TCA cycle equivalent of CO_2 and NADH. This multienzyme complex is very similar to the PDH complex in the intricacy of its protein makeup, cofactors and its mechanism of action. Also, as with the PDH complex, the reactions of the a-KGDH complex proceed with a large negative standard free energy change.

Although the a-KGDH of the complex is not subject to covalent modification, allosteric regulation is quite complex, with activity being regulated by energy charge, the NAD^+/NADH ratio and effector activity of substrates and products.

Succinyl-CoA and a-ketoglutarate are also important metabolites outside the TCA cycle. In particular,a-ketoglutarate represents a key anapleurotic metabolite linking the entry and exit of carbon atoms from the TCA cycle to pathways involved in amino acid metabolism. α-ketoglutarate is also important for driving the malate-aspartate shuttle. Succinyl-CoA, along with glycine, contributes all the carbon and nitrogen atoms required for the synthesis of protoporphyrin heme biosynthesis and for non-hepatic tissue utilization of ketone bodies.

SUCCINATE DEHYDROGENASE (SDH)

Succinate dehydrogenase catalyzes the oxidation of succinate to fumarate with the sequential reduction of enzyme-bound FAD and non-heme-iron. In mammalian cells the final electron acceptor is coenzyme Q_{10} (CoQ_{10}), a mobile carrier of reducing equivalents that is restricted by its lipophilic nature to the lipid phase of the mitochondrial membrane.

succinate — *succinate dehydrogenase* (FAD^+ → $FADH_2$) → fumarate

FUMARASE (FUMARATE HYDRATASE)

The fumarase-catalyzed reactions specific for the *trans* form of fumarate.

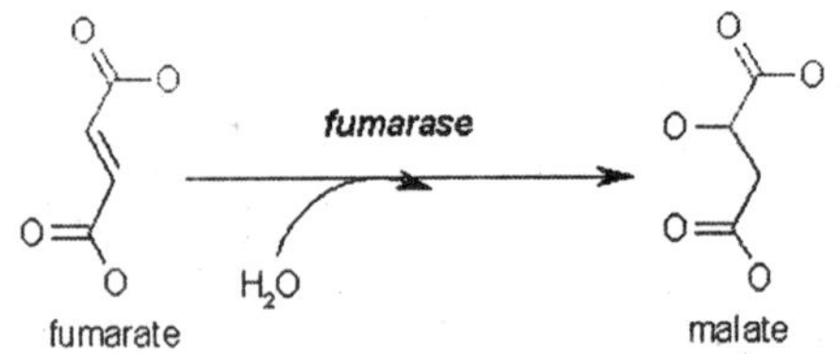

The result is that the hydration of fumarate proceeds stereospecifically with the production of L-malate.

MALATE DEHYDROGENASE (MDH)

L-malate is the specific substrate for MDH, the final enzyme of the TCA cycle. The forward reaction of the cycle, the oxidation of malate yields oxaloacetate(OAA).

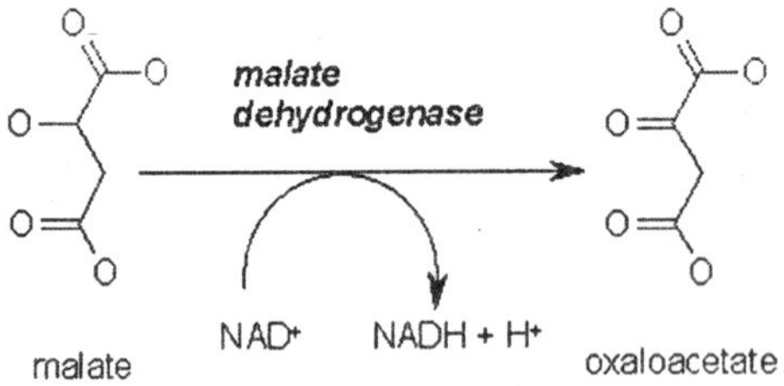

In the forward direction the reaction has a standard free energy of about +7 kcal/mol, indicating the very unfavourable nature of the forward direction. As noted earlier, the citrate synthase reaction that condenses oxaloacetate with acetyl-CoA has a standard free energy of about -8 kcal/mol and is responsible for pulling the MDH reaction in the forward direction. The overall change in standard free energy change is about -1 kcal/ mol for the conversion of malate to oxaloacetate and on to succinate.

The overall stoichiometry of the TCA cycle is:

acetyl-CoA + 3NAD+ + FAD + GDP + Pi + 2H2O →
2CO2 + 3NADH + FADH2 + GTP + 2H+ + HSCoA

REGULATION OF THE TCA CYCLE

Regulation of the TCA cycle. like that of glycolysis, occurs at both the level of entry of substrates into the cycle as well as at the key reactions of the cycle. Fuel enters the TCA cycle primarily as acetyl-CoA. The generation of acetyl-CoA from carbohydrates is, therefore, a major control point of the cycle. This is the reaction catalyzed by the PDH complex.

By way of review, the PDH complex is inhibited by acetyl-CoA and NADH and activated by non-acetylated CoA (CoASH) and NAD^+. The pyruvate dehydrogenase activities of the PDH complex are regulated by their state of phosphorylation. This modification is carried out by a specific kinase (PDH kinase) and the phosphates are removed by a specific phosphatase (PDH phosphatase). The phosphorylation of PDH inhibits its activity and, therefore, leads to decreased oxidation of pyruvate. PDH kinase is activated by NADH and acetyl-CoA and inhibited by pyruvate, ADP, CoASH, Ca^{2+} and Mg^{2+}. The PDH phosphatase, in contrast, is activated by Mg^{2+} and Ca^{2+}.

Since three reactions of the TCA cycle as well as PDH utilize NAD^+ as co-factor it is not difficult to understand why the cellular ratio of NAD^+/NADH has a major impact on the flux of carbon through the TCA cycle.

Substrate availability can also regulate TCA flux. This occurs at the *citrate synthase* reaction as a result of reduced availability of oxaloacetate. Product inhibition also controls the TCA flux, e.g. citrate inhibits *citrate synthase*, a-KGDH is inhibited by NADH and succinyl-CoA. The key enzymes of the TCA cycle are also regulated allosterically by Ca^{2+}, ATP and ADP.

THE ELECTRON TRANSPORT CHAIN

2NADH + 2H+ + O2 ==> 2NAD+ + 2H2O

The oxidation of NADH occurs in a stepwise manner involving an electron transport chain. Reducing equivalents go via a series of carriers to oxygen. Electrons flow downhill, energetically. Energy released in this process is used to generate the Proton Motive Force which is used by the cell to make ATP.

COMPONENTS OF THE ELECTRON TRANSPORT CHAIN (ETC)

The components of the ETC are membrane bound. Their

redox potentials vary between NAD (Eo = -320mV) and O2 (+820mV).

Note that the redox potentials of different cytochromes and flavoproteins etc. may differ even though they have the same prosthetic groups.

The surrounding protein modifies the Eo of the heme and flavin groups. The mitochondrial ETC is fixed in composition and is longer and generates more energy than that of *E. coli.* In contrast, many bacteria, including *E. coli* have multiple alternative ETCs produced in response to different environmental conditions.

NADH Dehydrogenase

Found in the cytoplasmic membrane with other components of the ETC. *E. coli* contains two alternative NADH dehydrogenases: NADH-DH I contains one FAD, several FeS groups and one bound ubiquinone(UQ) per enzyme.

This isoenzyme is coupled and generates PMF. Encoded by the gene *nuo* (NADH UQ oxidoreductase). Reducing equivalents go from NADH ϕ FAD ϕ FeS ϕ UQ. NADH-DH II has FAD & UQ but contains no FeS clusters and is not coupled and cannot generate PMF.

Encoded by *ndh.* It is used when there is surplus NADH to be oxidized but the cells have enough energy. The mitochondrial enzyme resembles NADH I except that it has FMN instead of FAD.

Ubiquinone

All ETCs contain quinones. Quinones are hydrogen carriers and carry 2H at a time. The structure of the quinone nucleus varies: ubiquinone (respiration) & plastoquinone (photosynthesis) each have a single aromatic ring whereas menaquinone (anaerobic respiration) has a double ring. Ubiquinone is sometimes called coenzyme Q and its side chain (R-group) is made of 6-10 isoprenoid (C5) units depending on the organism. In*E. coli* n = 8 and in mitochondria n = 10.

CH_3O ... CH_3 $\xrightarrow{2H}$ CH_3O ... CH_3

UBIQUINONE UBIQUINOL

There is more ubiquinone than all other ETCcomponents together. A small portion of the ubiquinone is tightly bound to ETC proteins and the rest is free to diffuse around in the membrane - the so-called ubiquinone pool.

The R-group = $-[CH_2-CH=C(CH_3)-CH_2]n-H$ I

Iron Sulfur Proteins (FeS Proteins)

These have iron but no heme group, instead the iron is linked to acid-labile, i.e. inorganic, sulfur. Sometimes called non-heme-iron (NHI) proteins. The FeS groups are electron carriers. Each FeS group carries only one electron even though it may have 2 or 4 iron atoms. The electron is shared among the irons:

$$e + Fe2+ = Fe3+$$

In *E. coli,* FeS groups are found in NADH-dehydrogenase-I and in the cytochrome b region of the chain. In mitochondria there are at least 7 FeS centres. Four are associated with mitochondrial NADH-dehydrogenase two are associated with cytochrome b and one with cytochrome c1.

Several variant structures, the most common are the flat Fe2S2 and the cube shaped Fe4S4 groups. Both are bound to the protein by four cysteine residues.

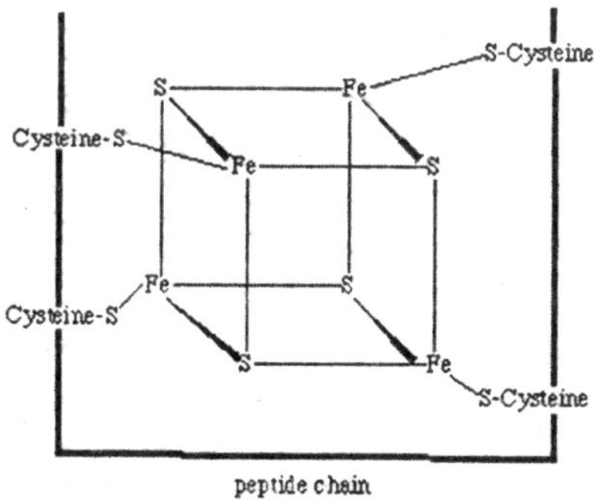

Fig. Peptide Chain

FLAVOPROTEINS

The NADH-dehydrogenases are flavoproteins. Several other flavoproteins join the ETC as side branches at ubiquinone, thus missing coupling site #1. Examples are succinate dehydrogenase(SDH), lactate dehydrogenase(LDH) and glycerol-P dehydrogenase.

Note that E. coli contains two types of LDH:

- Soluble, NAD-linked, converts pyruvate to lactate in fermentation
- Membrane-bound, FAD-linked, oxidizes lactate to pyruvate in air and feeds into ETC. There are actually two FAD-linked isoenzymes, one each for the D- & L- isomers of lactate.

Flavoproteins may also contain FeS groups (e.g., SDH, NADH-DH I). Usually FAD or FMN is not bound covalently, but occasionally FAD may be bonded to a histidine (e.g. in SDH). Flavoprotein dehydrogenases are linked to the ETC and are not reoxidised by molecular oxygen. In contrast, flavoprotein oxidases are not linked to the ETC and may be reoxidized directly by O2 e.g., aldehyde oxidase, glucose oxidase. Facultative anaerobes such as *E. coli* have few oxidase type flavoproteins.

CYTOCHROMES

Electron carriers with Fe-porphyrin (heme) groups. Alternate between Fe3+ and Fe2+. The Fe in heme has six coordination positions, four of which are filled by N-atoms of the heme. In most cytochromes both positions 5 and 6 are filled by amino acid residues and reaction with O2 is prevented. In hemoglobin position 5 is filled by a histidine of the protein and position 6 is free to bind O2. The same is true of the final cytochromes in the ETC which react with molecular oxygen. In mitochondria this is cyt a/a3, in *E. coli* both cyt d and cyt o are terminal oxidases. Cytochrome oxidases can bind CO, CN- or HS- (from H2S) instead of O2 and the enzyme activity is then killed.

In mitochondria and many obligately aerobic bacteria there is a full set of cytochromes: Electrons go from UQ to cyt b to cyt c1 to cyt c to cyt a/a3. Cyt a/a3 is known as cytochrome c oxidase since it oxidizes cyt c at the expense of molecular oxygen. Cyt a/a3 contains copper ions as well as heme. In facultative anaerobes there is a shorter chain and cyt c & cyt c1 are both missing. Electrons go from UQ to cyt b to cyt o or d.

In E. coli high aeration gives mostly the cytochrome o complex:

- Heme b (b555 & b562) and 2 copper ions.
  Low aeration results in appearance of the cytochrome d complex:
- Heme b (b558 & b595) and 2 heme d, but no copper.

The Km values for O2 are 0.2mM for cyt o and 0.02mM for cyt d which is consistent with the idea that cyt d is needed when the concentration of O2 is low. Thes are sometimes called ubiquinol oxidases since they oxidize UQ at the expense of molecular oxygen. There is no cytochrome c, c1 or a in the *E.coli* aerobic respiratory chain. Cyt d used to be called cyt a1 but is not really an a-type. A c-type cytochrome is found in the anaerobic nitrite reductase of *E.coli.*

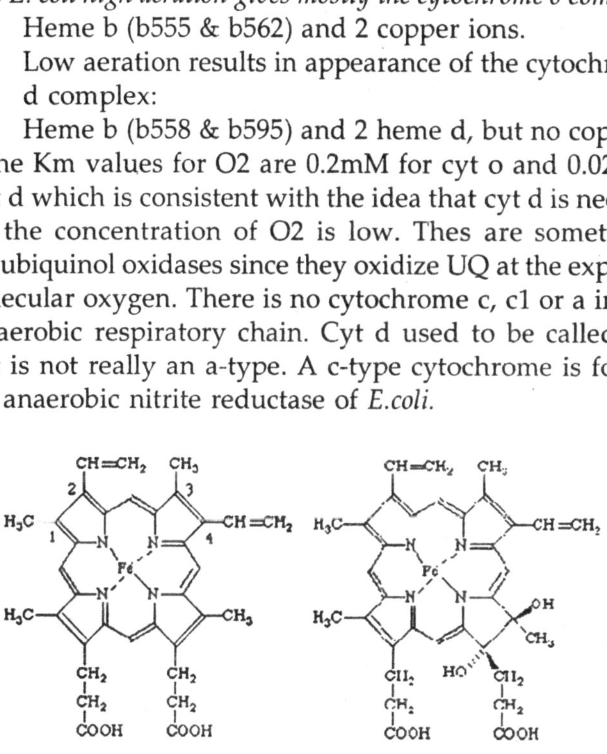

Protoheme IX = Heme b Heme d [a chlorin]

Porphyrin = Heme without the Fe atom and with 2 H atoms instead. The four 5-membered rings each containing N are pyrrole rings. Cytochromes are classified according to the substituents at positions 2 and 4 of the porphyrin/heme ring:

Cytochrome b: vinyl groups at both 2 and 4 positions. This gives protoheme IX and is also found in hemoglobin, myoglobin, catalase and peroxidase. Cytochrome o of *E. coli* is a b-type.

Cytochrome c: Positions 2 and 4 both are -CH(CH_3)-S-Protein.

Cytochrome a: Position 2 has a 17-carbon side chain, position 4 is vinyl and position 8 has -CHO (instead of the usual CH_3).

Chlorins are hemes in which one or more of the pyrrole rings are reduced. Heme d of *E.coli* cytochrome d (= cyt a1) and siroheme of sulfite reductase are chlorin derivatives.

THE PROTON MOTIVE FORCE

Originally, it was thought that ATP was made by direct chemical phosphorylation at 3 places along the ETC where sufficient energy was available from the drop in redox potential.

These are the three coupling sites. It is now known that 2 protons are expelled at each coupling site, so generating the Proton Motive Force (PMF). ATP is made indirectly using the PMF as a source of energy. Each pair of protons yields one ATP. In *E. coli* and many bacteria there is no cytochrome c or c1 and thus coupling site #2 is missing so that only 2 ATP per NADH may be made.

CHEMICAL COUPLING THEORY

Model is reactions such as glyceraldehyde phosphate dehydrogenase and pyruvate dehydrogenase where a redox reaction directly generates a high energy chemical intermediate, ie: diphosphoglyceric acid and acetyl-lipoate respectively. Disproven and of historical interest only.

CHEMIOSMOTIC THEORY

ATP synthesis is coupled indirectly to electron transport via the Proton Motive Force (PMF). In the chemiosmotic model, each of the coupling sites is responsible for extruding protons, so creating a proton concentration gradient across the membrane. The ATP synthase uses this gradient to energize the synthesis of ATP.

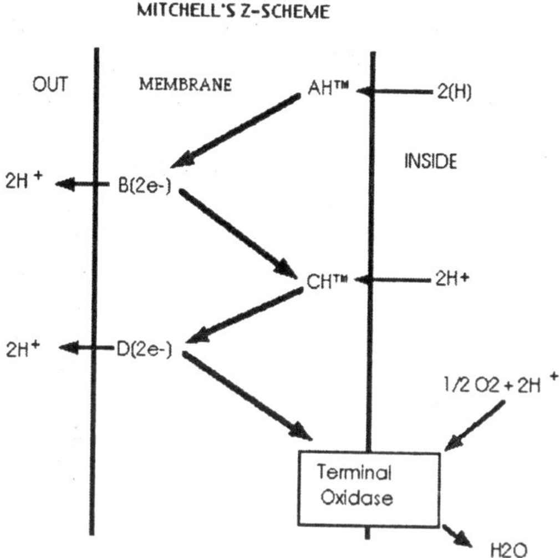

Fig. Mitchell's Z-scheme

The membrane must be impermeable to H+ (and OH-). Since H+ is pumped out, the exterior will become acidified and the cell interior should become alkaline relative to the outside. In practicc the internal pH is held constant at 7.5 approx. The proton gradient may be all or partially converted to a charge difference by exchanging potassium ions for protons. The PMF thus consists of two components: a pH difference (Delta pH) and a charge difference (or membrane potential, Delta Psi) across the membrane.

STOICHIOMETRY

Original theory of Mitchell has 2H+ extruded per site with 3 sites per mitochondrion or 2 sites per *E. coli* and hence P:O ratios of 3:1 or 2:1 assuming 2H+ can drive formation of one ATP. (P:O ratio = ATP per oxygen atom consumed = ATP per 2 electrons sent down ETC). Some data suggests 3 or 4 protons may be extruded at certain sites, but this only applies to mitochondria

MITCHELL'S Z-SCHEME

In Mitchell's original Z-scheme the ETC is arranged in

a series of loops with hydrogen carriers facing the inside of the bacterial cell or mitochondrion and electron carriers facing outwards.

Each hydrogen carrier receives two hydrogen atoms –2 and it transfers 2e- to the following electron carrier. Since 2 = 2H+ + 2e- this leaves two protons (2H+) to dispose of and these are expelled so generating the PMF.

The electron carrier transfers the electrons to the next hydrogen carrier. In order to get 2, two 2H+ must also be picked up from the inside of the bacterial cell. The ETC contains one proton extrusion loop for each coupling site.

MITCHELLS EQUATION

Proton chemical potential: μ = 2.3RT(pHi - pHo) = 2.3RT Delta pH

Electrical potential: m = F(Psi i- Psi o) = F DeltaPsi

i refers to inside, o to outside, F = Faradays constant, Psi = charge

Electrochemical potential = m(chemical) + m(electrical):

μ(H+) = F DeltaPsi - 2.3RT Delta pH.

Proton Motive Force:

PMF = μ(H+)/F = DeltaPsi - 60 Delta pH

PMF is the proton motive force in millivolts. Conversion factor is approx. 60mV/pH unit at 37°C. Note that the relative contributions of DeltaPsi and Delta pH to PMF vary with the pH of the growth medium. Note also that PMF can only be maintained as long as membrane forms a sealed vesicle.

To generate 100mV of PMF requires the extrusion of about 60,000 protons/cell. The PMF generated by *E. coli* is 150-200 mV. How the intracellular pH is maintained at pH7.5 is unknown. It is known that at an external pH of 7.5 the Delta pH is replaced by a DeltaPsi by swapping K+ for H+.

In effect, *E. coli* is a tiny living battery. To convert ADP to ATP requires 7.3 kcal/mole energy input. This is equivalent to the energy stored in a pair of protons at a pH gradient of 3.5 pH units or at a charge difference of 240mV. Uncouplers are proton ionophores(= ion carriers).

They carry H+ across membranes and consequently allow

the PMF to collapse. Both the protonated and unprotonated forms can diffuse.

Protons are picked up on the outside and released on the inside. eg: 2, 4-Dinitrophenol (2,4-DNP), carbonyl cyanide-p-trifluoromethoxyphenylhydrazone (FCCP) and the similar chloro compound CCCP.

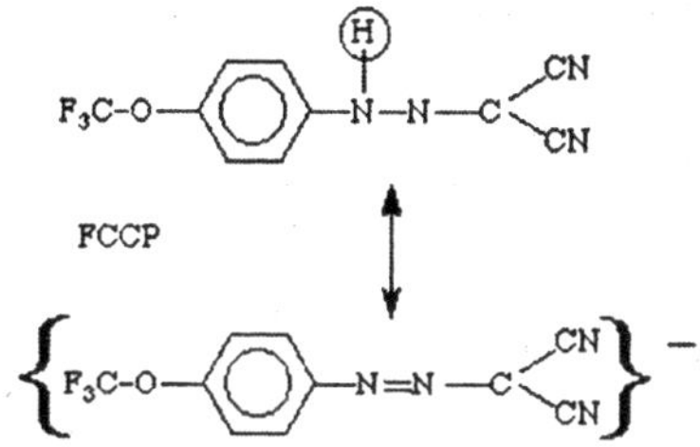

FCCP

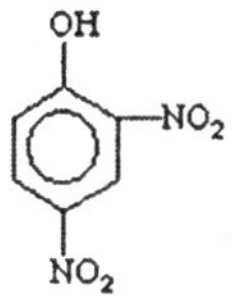

2,4 - DNP

Other Ionophores

Carriers exist for ions other than the proton eg:

- Valinomycin carries potassium(K+) or rubidium(Rb+) ions and will therefore collapse the DeltaPsi but not the Delta pH
- Nigericin exchanges Na+, K+ or Rb+ for H+ so collapsing Delta pH but not DeltaPsi ie opposite effect to valinomycin
- Gramicidin is not a carrier but forms a channel in the membrane allowing through H+, Na+, K+, & Rb+ thus collapsing both Delta pH & DeltaPsi

Measurements of DeltaPsi and Delta

Delta pH may be measured by the distribution of weak acids across the membrane. This depends on the Delta pH. Acids used eg: acetic acid, salicylic acid, DMO (5,5-dimethyloxazolidine-2,4-dione):

- Membrane permeable to undissociated acid(HA) but not to H+ or A-
- On each side of membrane HA ∞ H+ + A-
- The extent of dissociation depends on pH
- Thus extent of dissociation and hence the amount of HA differs on either side of membrane and this can be measured.

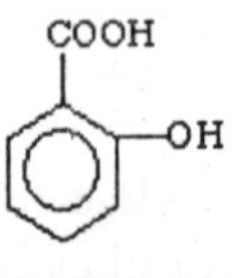

SALICYLATE 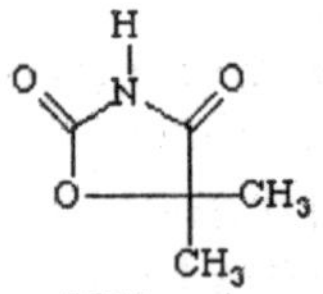 DMO 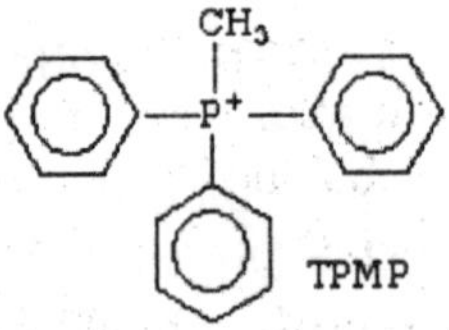 TPMP

DeltaPsi may be measured indirectly by a similar method. A membrane permeable cation will be distributed according to DeltaPsi. For example, the ionophore valinomycin plus K+ or Rb+, or a lipophilic cation such as TPMP+ (triphenylmethyl-phosphonium) can be used. DeltaPsi may be measured directly with a voltmetre and electrodes:

- Grow *E. coli* with mecillinam, a penicillin derivative which, in sublethal amounts, produces giant spherical cells (upto 6 microns diam).
- Use small electrodes! (0.08 to 0.2 microns at tip).
- Coat electrodes with phospholipid to get a good seal with the cell membrane - required to avoid PMF collapsing.
- Read DeltaPsi from voltmetre.

Table. Example Proton Motive Force Values

Organism	pHo	-60DeltapH	Delta Psi	PMF
Fermentative				
Staphylococcus lactis	5.0	58 (benzoate)	95 (dye)	158
Staphylococcus faecalis		5.0	54 (DMO)	1 7 0
(DDA)	224			
Fermentative /Respiratory				
Bacillus subtilis	5.0	120 (salicylate)	30 (TPMP)	150
Bacillus subtilis	7.5	0 (salicylate)	140 (TPMP)	140
Escherichia coli	5.5	72 (DMO)	57 (Rb+/Val)	129
Escherichia coli	7.0	38 (DMO)	122 (Rb+/Val)	160
Escherichia coli	5.5	120 (DMO)	100 (electrodes)	220
Escherichia coli	7.5	0 (DMO)	142 (electrodes)	142
Photosynthetic				
Rhodospirillum rubrum chromatophores	72	50(SCN-)	122	

as these are inside out, polarity of PMF is reversed		(CH3NH4+)		
whole cell vesicles		36 (acetate)	70 (TPMP)	106
Halobacterium halobium	6.0	73 (DMO)	113 (TPMP)	186
Halobacterium halobium	8.0	8 (DMO)	151 (TPMP)	159

OXIDATIVE PHOSPHORYLATION

ATP synthesis is driven by the PMF. It can be demonstrated that only the PMF and the ATP synthase enzyme complex are required in the presence of sealed membrane vesicles. Note: Experiment is usually done "inside outΔ with the ATPase stuck to the outside.

- Make liposomes (i.e., lipid vesicles)
- Add purified ATPase and ADP and Pi
- Add excess K+ plus Valinomycin to outside
- Positive charge builds up on inside - an artificial PMF
- The artificial DeltaPsi repels protons, derived from: H2O = H+ + OH-
- Protons exit via ATP synthase and ATP is made

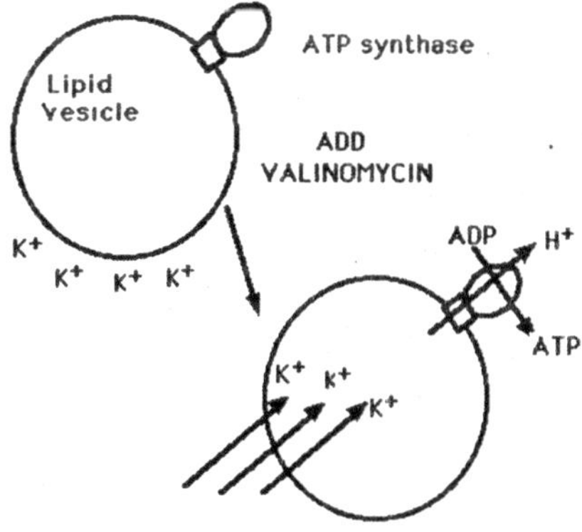

Fig. ATP synthase

ATPase and ATP synthase refer to the same enzyme - which direction the reaction goes depends on the conditions.

The ATP synthase consists of the headpiece, or F1-complex, which sticks into the interior of the cell and the

Fo-complex which is membrane-bound. The catalytic a and b subunits of the F1-complex are responsible for synthesizing ATP. The Fo complex consists of 3 proteins a, b and c (sometimes referred to as c, y, w). Fo forms a proton channel through the membrane.

Protein c is very hydrophobic (a "proteolipidΔ) and probably acts as the channel while a and b control assembly and binding to the F1-headpiece.

Fo alone makes membranes permeable to protons. Addition of F1 stops proton transport except to drive ATP synthesis. DCCD (dicyclohexyl carbodiimide) binds covalently to protein c and blocks the proton channel.

Oligomycin inhibits the ATPase reaction of the catalytic F1 complex. Oligomycin halts the whole electron transport chain & ATP synthetic pathway when added alone. However, if both oligomycin and an uncoupler are added then the ETC runs but no ATP synthesis occurs.

PROTEINS OF THE ATP SYNTHETASE

Protein Ratio Gene Subunit MW Function:

- α 3 uncA F1 55kd catalytic
- a (or c) 1 uncB Fo 30kd binds to F1
- ε 1 uncC F1 15kd regulation
- β 3 uncD F1 50kd catalytic
- c (or w) 10 uncE Fo 8kd proton channel
- b (or y) 2 uncF Fo 17kd binds to F1
- γ 1 uncG F1 31kd structure of F1
- δ 1 uncH F1 20kd binds F1 to membrane

The precise mechanism of ATP synthesis is obscure. The simplest scheme for ATP synthesis postulates removal of an oxygen from inorganic phosphate by two protons from the outside thus producing water on the outside and ATP on the inside.

The inorganic phosphate is closest to the proton channel with the ADP to the inside An alternative is the use of energy from the protons to alter the conformation of the catalytic a and β subunits. The high energy state then releases its energy by converting ADP to ATP.

ATP VERSUS THE FORCE

The ATPase can interconvert PMF and ATP. In the absence of a PMF the ATPase will hydrolyse ATP and generate PMF. Hence ATP and PMF are interconvertible. If an uncoupler is added to *E. coli* or a mitochondrion the PMF is dissipated. The ATPase then tries to re-create the PMF by hydrolyzing all the ATP. Oligomycin inhibits the ATPase and will prevent hydrolysis of ATP under these conditions just as it prevents synthesis of ATP from PMF.

ATP is used for biosynthesis and also drives certain transport systems. Arsenate inhibits ATP driven transport systems, but not those energized directly by the proton motive force. In eukaryotes the PMF is generated in the mitochomdria or chloroplasts and is all converted to ATP. In bacteria the PMF is used directly to energize several membrane bound operations:

PMF DRIVEN TRANSPORT SYSTEMS

- Symport: e.g., Lactose Substrate and H+ enter together.
- Antiport: e.g., Na+ H+ taken in and Na+ excreted.
- Complex: e.g., Melibiose Overall Melibiose plus H+ enter:
- Melibiose plus Na+ enter together ii) Na+ out/H+ in.

Energy Linked Transhydrogenase

Normally NADH is converted to ATP whereas NADPH is used as reducing power in biosynthesis. NADPH may be generated directly by certain metabolic pathways which generate NADPH instead of NADH.

A proportion of the glucose (around 20%) goes via the pentose phosphate pathway instead of the Embden-Meyerhof pathway. In many bacteria the isocitrate dehydrogenase in the Krebs cycle also makes NADPH. Another source is the NADP linked malic enzyme.

However if their is a shortage of NADPH then it is possible to convert NADH to NADPH. This requires energy in order to shift the equilibrium in favour of NADPH. Energy linked transhydrogenase is energized directly by the PMF and converts NADH + NADP to NADPH + NAD. Mechanism unknown. It is particularly important in some photosynthetic organisms which do not have any glucose to send down the pentose pathway. It is present in *E. coli* but not important.

Flagellar Motion

E. coli is powered by a proton drive - just like the ships of the galactic empire The M-ring of the basal body is located in the cytoplasmic membrane and rotates relative to the S- or stator ring of the basal body of the flagellum which is embedded in the cell wall.

Proton acceptor groups (e.g., -NH2) are present on the rotor-M-ring and anionic groups (e.g., COO-) are located over the exit channels for protons . Each time one H+ enters the input channel the ring turns a fraction of the revolution as the newly formed -NH3+ group is attracted towards the -COO- group.

ANAEROBIC RESPIRATION

Oxygen is the ultimate electron acceptor for the ETC when bacteria grow aerobically. However, bacteria can respire in the absence of oxygen if some other suitable oxidant is available to act as electron acceptor. Use of such alternative electron acceptors is known as anaerobic respiration.

It requires a modified ETC in order to transport electrons to the new acceptor. *E. coli* can use nitrate (NO3-), amine oxides (R3NO), sulfoxides (R2SO), fumarate and nitrite (NO2). Many sulfur compounds can be used by a variety of bacteria. "Sulfide fermentationΔ is not actually fermentation but is an example of anaerobic respiration with sulfate (SO42-) as the electron acceptor. "Methane fermentationΔ is also anaerobic respiration, in this case with carbon dioxide (CO_2) as the electron acceptor.

Do not confuse anaerobic respiration with the oxidative metabolism of lithotrophs such as *Nitrobacter*. Oxygen has a more positive Eo than nitrite, therefore oxidation of nitrite by oxygen produces energy. Nitrate has a more positive Eo than 2 from NADH or formate.

Therefore oxidation of NADH or formate by nitrate produces energy. Lithotrophs oxidize inorganic materials using O_2.

A substance with an Eo between O_2 and NADH may be used either to oxidize NADH or to reduce O2. e.g. the NO3-/NO2- couple which has Eo of +420mV compared to +820mV for O2/H2O and -320mV for NAD+/NADH.

ANAEROBIC NITRATE RESPIRATION: $NO_3 + 2 = NO_{2-} + H_2O$

Nitrite oxidation by *Nitrobacter* $NO_{2-} + 1/2O_2 = NO_{3-}$

NITRATE REDUCTASE

Nitrate reductase (NAR) of *E. coli* is a very large membrane bound protein (Total MW approx. 800,000). It is induced by nitrate and repressed by oxygen. When maximally induced it comprises upto 25% of the protein of the cytoplasmic membrane. Pyruvate formate lyase is present in anaerobic cells growing on nitrate and produces a lot of formate plus acetyl-CoA. Most of the acetyl-CoA is converted to acetate which is excreted. This is because the Krebs cycle is not complete during anaerobic growth because oxoglutarate dehydrogenase is not made. The formate is oxidized via formate dehydrogenase to give 2 plus CO_2. The reducing equivalents travel via the cytochrome b of the FDH complex to a quinone. NADH produced in glycolysis can also be reoxidized by nitrate. The quinone used anaerobically is mostly menaquinone (Vitamin K2). However, when grown on nitrate, which is a good oxidant, a mixture of ubiquinone and menaquinone is found.

Nitrate reductase contains 4 subunits:

- NarG protein x4 (= a) MW = 150,000, carries Mo & FeS groups
- NarH protein x4 (= b) MW = 60,000, carries FeS groups

- NarI protein (= g) cytochrome b, often lost on purification
- NarJ protein membrane binding subunit

The Mo is present as part of the low molecular weight (approx 1000) molybdopterin cofactor. All Mo proteins share the same cofactor, (except for nitrogenase which has a different Mo/Fe cofactor).

THE RESPIRATORY CHAIN TO NITRATE

Formate dehydrogenase is another large membrane protein. It contains molybdopterin, FeS groups and selenium. There are two types of FDH. The anaerobic respiratory FDH is found in anaerobic cells using nitrate or other alternative electron acceptors. The fermentative FDH is linked to hydrogenase and is produced anaerobically in the absence of alternative electron acceptors. The Se is present as selenocysteine ie cysteine in which sulfur has been replaced by selenium.

Most anaerobic reductases such as those for TMAO (trimethylamine oxide), DMSO (dimethyl sulfoxide) etc are molybdoproteins. Tungsten (W) is an analog of Mo. In the presence of tungstate (WO¢¤-) the processing of molybdate (MoO¢¤-) into molybdopterin is inhibited and nitrate reductase and other Mo enzymes are inactivated. Excess molybdate will remedy this i.e., tungstate is a competitive inhibitor.

The two anaerobic reductases which do not contain molybdenum are nitrite reductase - a heme protein and fumarate reductase - a flavoprotein.

Fumarate Reductase (FRD) FRD is repressed by both oxygen and nitrate but induced by fumarate. FRD contains FAD and FeS groups. There are 4 subunits. A large flavoprotein, a medium sized FeS protein and two small membrane anchor proteins.

FRD is connected via a cytochrome b to menaquinone. FRD cannot use ubiquinone, it has to use menaquinone. No molydenum or selenium are found in FRD. Note that fumarate is an organic molecule and the 2 go to saturate a double bond i.e., no water is formed as with nitrate, TMAO or DMSO etc.

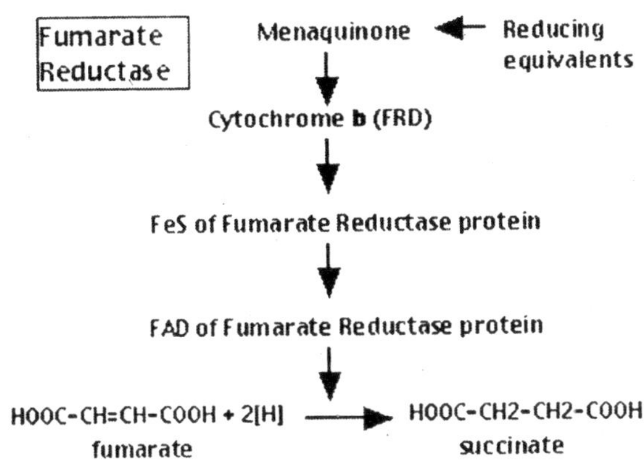

Fig. Respiratory Chain to Nitrate

Unlike O_2 and nitrate which both generate two pairs of protons per 2e- which travel down the electron transport chain and hence yield 2ATP, fumarate gives only one pair of protons and only one ATP.

When a cell changes from oxygen to fumarate as electron acceptor the PMF drops from around -200mV to approx. -100mV.

OXIDATIVE PHOSPHORYLATION

The NADH and $FADH_2$ formed in glycolysis, fatty acid oxidation and the citric acid cycle are energy-rich molecules because each contains a pair of electrons having a high transfer potential.

When these electrons are used to reduce molecular oxygen to water, a large amount of free energy is liberated, which can be used to generate ATP. *Oxidative phosphorylation is the process in which ATP is formed as a result of the transfer of electrons from NADH or $FADH_2$ to O_2 by a series of electron carriers.*

This process, which takes place in mitochondria, is the major source of ATP in aerobic organisms. For example, oxidative phosphorylation generates 26 of the 30 molecules of ATP that are formed when glucose is completely oxidized to CO_2 and H_2O. Oxidative phosphorylation is conceptually simple and mechanistically complex. Indeed, the unraveling of the mechanism of oxidative phosphorylation has been one of the most challenging problems of biochemistry.

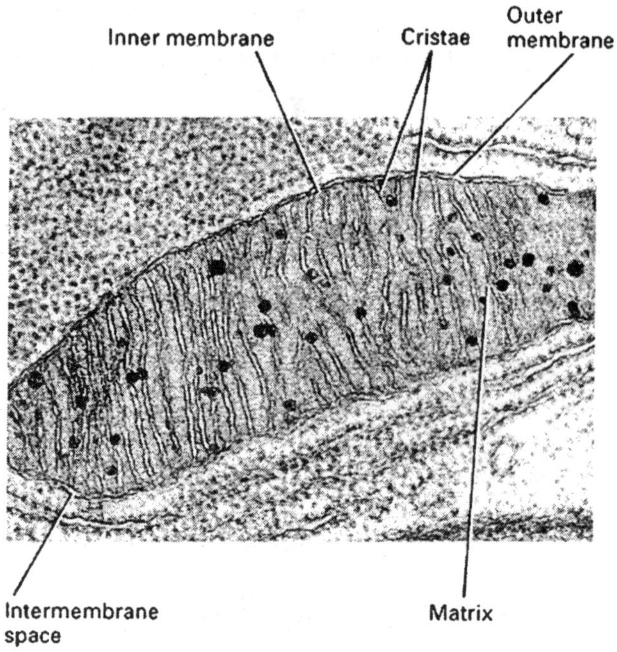

Fig. Electron Micrograph of a Mitochondrion

The flow of electrons from NADH or $FADH_2$ to O_2 through protein complexes located in the mitochondrial inner membrane leads to the pumping of protons out of the mitochondrial matrix.

The resulting uneven distribution of protons generates a pH gradient and a transmembrane electrical potential that creates a *proton-motive force.* ATP is synthesized when protons flow back to the mitochondrial matrix through an enzyme complex.

Thus, *the oxidation of fuels and the phosphorylation of ADP are coupled by a proton gradient across the inner mitochondrial membrane. Oxidative phosphorylation is the culmination of a series of energy transformations* that are called *cellular respiration* or simply *respiration* in their entirety.

First, carbon fuels are oxidized in the citric acid cycle to yield electrons with high transfer potential. Then, this electron-motive force is converted into a proton-motive force and, finally, the proton-motive force is converted into phosphoryl transfer potential.

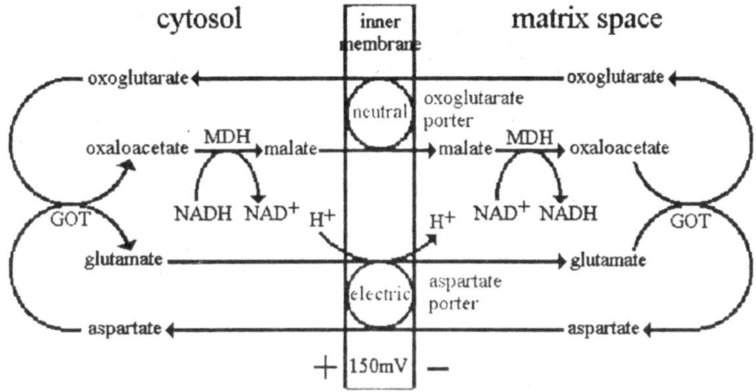

Fig. Essence of Oxidative Phosphorylation.

The conversion of electron-motive force into proton-motive force is carried out by three electron-driven proton pumps—NADH-Q oxidoreductase, Q-cytochrome *c* oxidoreductase and cytochrome *c* oxidase.

These large transmembrane complexes contain multiple oxidation-reduction centres, including quinones, flavins, iron-sulfur clusters, hemes and copper ions. The final phase of oxidative phosphorylation is carried out by *ATP synthase,* an ATP-synthesizing assembly that is driven by the flow of protons back into the mitochondrial matrix.

GLYOXYLATE CYCLE

It was stated above that fatty acids cannot be utilized for gluconeogenesis. Also, it was mentioned earlier that fatty acids are degraded via acetyl-CoA. Since acetyl-CoA is degraded via the TCA and TCA intermediates can be utilized for gluconeogenesis, why is it not possible to turn fatty acids into glucose?

This is because of the different roles of the TCA intermediates in the two processes:

- In the degradation of acetyl-CoA, they have a catalytic role – they are required for the cycle to run, but there is no net gain or loss of intermediates. Acetyl-CoA fed into the TCA is consumed without increasing the pool of TCA intermediates.
- In gluconeogenesis, oxaloacetate is drained from the pool of TCA intermediates. Thus, the TCA can only sustain gluconeogenesis to the extent it is supplied with new intermediates from other sources.

To turn fatty acids into glucose, we would therefore need a means to use acetyl-CoA for increasing the pool of intermediates.

Such a pathway does not exist in mammals, but it does exist in plants; it is known as the glyoxylate cycle. It is a 'side-road' to the TCA that allows them to use two molecules of acetyl-CoA per cycle to bring about the net synthesis of one C_4-intermediate.

Two reactions are required for this cycle:

- Isocitrate is split into succinate and glyoxylate by isocitrate lyase. Since the isocitrate dehydrogenase and the á-ketoglutarate dehydrogenase reactions are bypassed, the loss of two carbons as CO_2 is avoided; these carbons are retained in the form of glyoxylate.
- Glyoxylate combines with an 'extra' acetyl-CoA to form one molecule of malate. This reaction is catalysed by malate synthase and like the citrate synthase reaction it is pushed forward by the concomitant hydrolysis of coenzyme A.

We know that many plant seeds are very rich in oil (=fat). The glyoxylate cycle enables plant seeds to store metabolic energy and carbon as fat and to use it for the synthesis of glucose and other carbohydrates during germination.

Fat is water-free and has approximately twice the content of metabolic energy per gram of carbohydrates; it is therefore both more compact and more resistant to microbial degradation than starch.

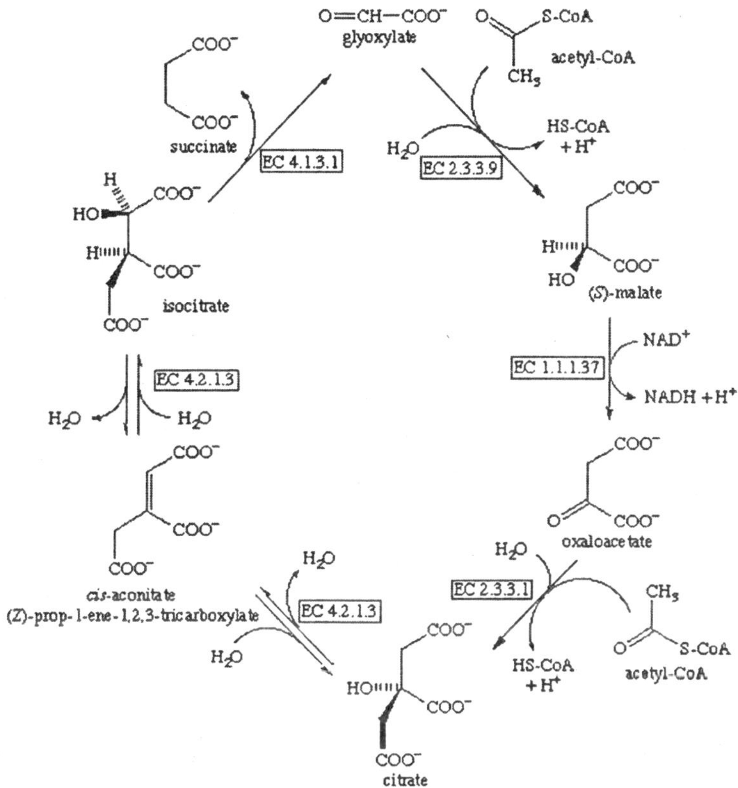

Fig. Glyoxylate Cycle

Chapter 6

Bacteria

BACTERIAL GROWTH

Bacteria are widely present in soil, water, dust, air, human contamination sources and animal reservoirs. Many are accidental intruders into human processes. The basic differences between prokaryotic cells and eukaryotic cells. A complete understanding of antimicrobials and their mechanism of action is critical for mastery of medical microbiology.

Structure and Growth: Prokaryotes vs Eukaryotes		
Prokaryotes		**Eukaryotes**
Nucleoid	>>>>>>	Nucleus
Operon Gene Structure		Inron & Exon Gene Structure
Free Polysomes	>>>>>>	Ribosomes assoc. with Rough E.R.
Polycistronic mRNA		Monocistronic mRNA
No E.R.	>>>>>>	Rough & Smooth
Free Enymes	>>>>>>	Lysosomes
Proton Motive Force (pmf)	>>>>>>	Mitochondria
n-formyl-Met >>>>>> protein modification		None

The purpose for categorizing bacteria is to establish phylogeny; genetic and evolutionary relationships and to devise chemotherapy strategies based on better comprehension. Bacterial morphology, biochemistry and physiology (autotrophs vs heterotrophs; aerobic vs anaerobic; culture/survival conditions) and sequence homology (moleculary biology) is the basis for bacterial diversity. Diagnostics is a competitive market built on exploiting these differences among the bacteria. Diagnostics uses rapid integration of new technologies for state of the art, fast, simple medical care to benefit patients. Major concerns include minimizing cost while maximizing reliability and reproducibility.

All bacteria have a lipid bilayer cytoplasmic membrane with embedded proteins that control movement of nutrients and waste products across the cell boundary. Gram-negative cells have a second, outer membrane (OM) that is very different from the cytoplasmic membrane, with porin proteins that limit access of large molecular weight substances to the cell interior. Also present in the outer part of the OM bilayer is lipopolysaccharide (LPS). The OM encloses another unique Gram-negative feature, the periplasmic space where a thin peptidoglycan layer (~3nm thick) and enzymes including the penicillin-binding proteins are located.

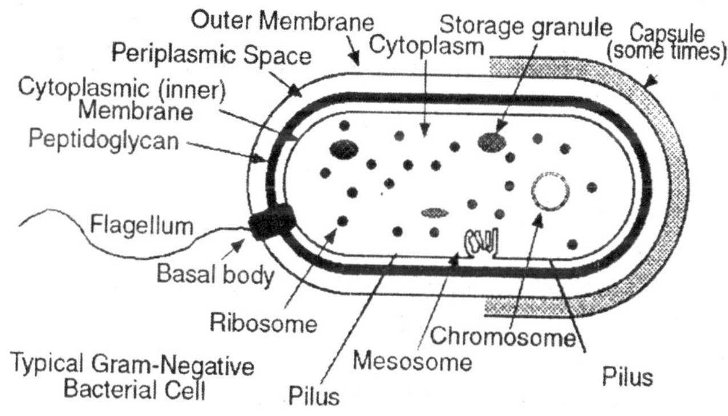

Fig. Structure of Bacteria

Peptidoglycan is a substance made exclusively by bacteria using repeating N-acetyl-D-muramic acid (NAM) and N-acetyl-D-glucosamine (NAG) that is highly crosslinked. Peptidoglycan provides the cell with mechanical strength and resistance to osmotic changes. There is a peptidoglycan layer in Gram-positive bacteria also, but it is much more substantial (10-80nm thick) and more highly cross-linked. Gram-positive bacteria also have teichoic and lipoteichoic acids that set them apart from Gram-negative cells. Bacteria from both groups may have a loose capsule outside the cell wall and the capsule composition is characteristic of the species. Capsules form a protective barrier that makes it difficult for the immune system to r ecognize and opsonize the bacterial cell.

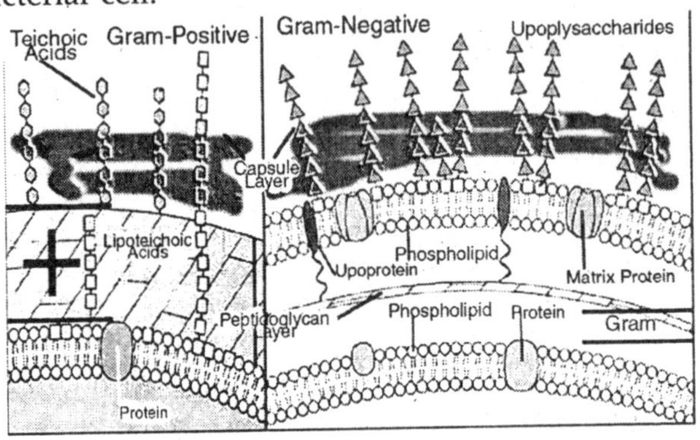

Fig. Gram Positive and Gram Negative

BACTERIAL STAINING

The first step for any bacterial stain is heat fixing the bacterium to the microscope slide. Next, a stain is used to provide contrast between the microbe and the background. Simple stains like the Methylene Blue Stain are taken up by most bacteria and allow visualization without differentiating among the various types of bacteria.

Other stains like the Gram stain and Acid-fast stain allow different staining patterns with the various types of bacteria. Bacteria can be grouped according to Gram stain and light microscopic appearance as either Gram-positive or Gram-negative. The procedure uses crystal violet primary stain with iodine as a mordant. In the second step, Gram-positive bacteria are resistant to decolourization with alcohol and remain a purple/dark blue colour after treatment.

The opposite is true for the Gram-negative bacteria that are decolourized during the alcohol treatment. Gram-negative bacteria show instead the pink colour of the safranin counterstain used in the final staining step. The difference in staining pattern is due to a major distinction in the bacterial cell wall. Some bacteria cannot be stained using the Gram-stain method. The spirochetes (Borrelia, Leptospira and Treponema) do not stain well, but have the Gram-negative cell wall structure.

These bacteria can be visualized using silver staining or dark-field microscopy. Chlamydia and Rickettsia species have unconventional cell walls and must be stained using special methods and Mycoplasma species don't have a cell wall at all. Mycobacterium species are essentially Gram-positive, but have a thick, waxy coating and can only be detected using the Acid-Fast Stain (carbol fuchsin). Bacteria that move about have motility provided by flagellae (singular, flagella) that extend from the cell surface and reach beyond the capsule.

Flagellar ultrastructure shows differences in membrane anchoring between the bacterial groups. Bacterial pili (singular pilus, also known as a fimbria) are small, hair-like structures that mediate adherence to target host structures. Pili are made of adhesin proteins and some bacteria possess the F′ or sex pilus that allows exchange of DNA with "receptive" cells in a process known as conjugation.

BACTERIAL SHAPES

Cocci (= round; prototype Staphylococcus). Bacilli (= rods; prototype Escherichia coli).

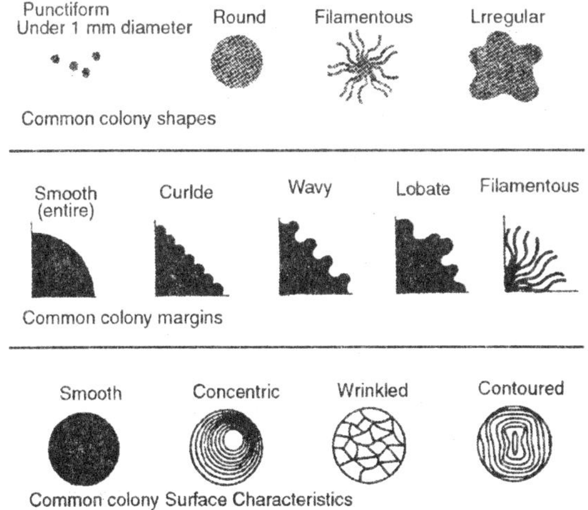

Fig. Shape of Bacteria

ANTIMICROBIAL SITES OF ACTION

Cell wall synthesis inhibitors are a major class of antibiotics, including penicillins and penicillin derivative antibiotics (amoxicillin, oxacillin, etc), the cephalosporins and vancomycin (a peptide antibiotic). The penicillin and cephalosporin drugs localize to the periplasmic space in Gram-negative bacteria, where the bacterial enzymes that synthesize peptidoglycan are found (transpeptidases that crosslink PG strands). This is also where the ^{2}-lactamase enzymes that allow antibiotic resistance are located. Drugs having the ^{2}-lactam ring structure are susceptible to hydrolysis at the highly constrained ring. Cephalosporins have an additional five-membered ring and represent more complex forms of ^{2}-lactam class antibiotics.

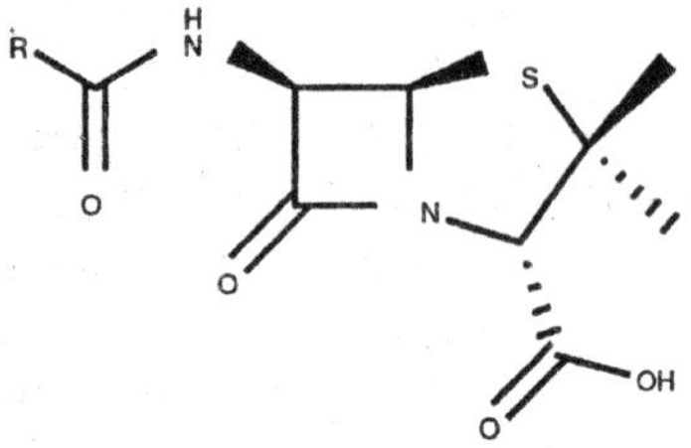

Fig. Structure of Penicilline

Vancomycin targets cell wall cross linking as well, binding to the ends of the PG peptide pentamer and sterically hindering cross-linking across PG strands. It is a peptide antibiotic. Because it works at a different site than the penicillin/cephalosporins, vancomycin is very effective against Gram-positive organisms and it is not susceptible to bacterial ^{2}-lactamase enzymes.

Vancomycin is reserved for renal dialysis patients in many formularies, in order to retain its effectiveness for this highly susceptible patient population. Nutrient mimic antibiotics like folic acid metabolism inhibitors include the sulfonamides and trimethoprim. They are similar to nutrient molecules and this facilitates their uptake.

Cell membrane de-energizing antibiotics like the polymyxins are effective against Grampositives more than

Gram-negatives. It is thought that the Gram-negative outer membrane effectively inhibits penetration of these large antibiotics into the interior where the cytoplasmic membrane could be attacked. Nucleic acid synthesis. DNA synthesis is another very effective sight of antibacterial inhibition. DNA gyrase is the target (fluoroquinolones and nalidixic acid are the drugs). For RNA synthesis, DNA-directed RNA polymerase is the target bacterial molecule and Rifampin (rifampicin) is the drug.

These agents are considered to be bactericidal because they inhibit fundamental cellular processes. Protein synthesis is the antibiotic site of action for the remainder of the antibacterial compounds. There are 50S ribosomal subunit inhibitors (like chloramphenicol and erythromycin) and 30S ribosomal subunit inhibitors (streptomycin, gentamicin) and tRNA binding inhibitiors (tetracycline; also binds to the 30S ribosomal subunit).

BACTERIAL RESISTANCE FACTORS	
Structure	**Effects on Antibiotics**
GRAM-NEGATIVES	
LPS of Gram negative	Retards/prevents penetration of bulky, high MW antibiotics (novobiocin, erythromycin).
Lipid bilayer of Gram-negative outer membrane	Penetration of water-soluble drugs is severely hindered.
Hydrophilic pores of Gram-negatives	Allow penetration of water- solubles up to 650 Daltons (sulfonamides)
Nutrient receptor transport proteins of Gram-negative outer membrane	Agents related to nutrient transport (sideromycins) utilize these natural receptors
GRAM-POSITIVES	
Teichoic and teichuronic acids	Strong anionic character of these polymers may affect rate of penetration
ALL BACTERIA	
Nutrient transport proteins of cytoplasmic membrane	Facilitate rapid penetration of agents related to nutrients by structure (D-cycloserine, phosphomycin; tetracyclines)
Mutations in PBPs Absence of high-affinity PBPs	Changed susceptibility to penicillin and derivative β-lactam antibiotics
β-lactamase enzymes	Target site is the β-lactam ring in penicillin and similar antibiotics (β-lactam ring)

BACTERIAL GROWTH

The definition of bacterial growth is increase in biomass.

For growth in liquid media, a growth cycle determination can be made. According to the graph (right), lag phase is the horizontal beginning period when the enzymes are being synthesized that will accomplish the bacterial utilization of available nutrients.

In the stationary phase, there is no net increase in biomass, but rather a steady state where dead cells are quickly replaced with newly synthesized cells. The entire cycle is subject to chemostasis and death phase occurs when nutrients are depleted and toxic metabolic by-products build up to inhibit new growth.

Growth of bacteria on solid media or agar allows for enumeration (determination of viable cell number), selection of isolated colonies and determination of colony morphology (many bacteria have characteristic colony appearances). Pure cultures are obtained when a single kind of bacteria is present.

BACTERIAL DEATH

Death is defined as the loss of viability and loss of ability to produce progeny. Sterilization is used when the inactivation of all microbes is an absolute requirement, such as the sterilization of surgical instruments. Sanitizing a surface results in the reduction of bacterial content and antisepsis is the sanitizing of a body surface.

Sterilization is the removal of all microorganisms while disinfection is the removal of vegetative cells only. Viability is the capability to grow and develop as an independent unit and lethality is irreparable damage to the genome, envelope, or the proteins required for protein synthesis.

CATEGORIES OF INFECTION

Physical agents of microbial death include moist heat, autoclaving, dry heat pasteurization and freezing or lyophilization. Moist heat is efficient at protein denaturation. Autoclaving is moist heat intensified by pressure. By definition, autoclaving is 121°C (250°F) for 15-20 minutes at 15 psi pressure.

Dry heat must be applied for 1-2 hours at 160oC to

effectively oxidize microbes directly. Both these methods result in complete microbial killing, including spores. Pasteurization is the heating of a liquid to 62oC for 30 minutes and it kills known milk pathogens without completely sterilizing the milk (Lactobacillus species are not killed). Freezing causes microbial death by concentration of solutes. Bacterial cells can be preserved by lyophilization (freeze-drying) or by adding an emulsifying agent.

HIGH RISK **Surgical Instruments** **Needles, syringes** **Human tissues**	▪ Introduction into sterile body areas ▪ Close contact with mucous membrane or damaged skin ▪ Disposal of infectious waste materials	**STERILIZATION** Dry Heat Moist Heat Autoclaving Irradiation Chemicals (ethylene oxide) Filtration (physical removal)
MEDIUM RISK **Respiration Devices** **Endoscopes** **Thermometers**	▪ Contact with mucous membranes ▪ Prior to use with immunocompromised ▪ Contaminated with pathogenic material	**DISINFECTION** Alcohol Chlorine solution Phenolics Hydrogen peroxide
LOW RISK **Wheel chairs** **Patient Gurneys, Beds** **Walls, Floors, Sinks**	▪ Mainly in contact with healthy skin ▪ Little or no actual patient contact	**SANITIZING (CLEANING)** Surfactants Quaternary ammonium compounds

Mechanical agents of microbial death include ultrasonic treatment, filtration (<400 nm filter pore size) or pressure. Chemical agents of microbial death include acid or alkali solutions (pH extreme), salts, heavy metals (silver nitrate), halogens (chlorine), alkylating agents (ethylene oxide; formaldehyde – these cause protein and nucleic acid damage), surfactants (detergents that compromise cytoplasmic membrane) and phenol and alcohols that cause protein denaturation.

Chemical antimicrobial modes of action include physical denaturation and inactivation of proteins; de-energizing of the cell membrane and disruption of cellular membrane permeability; and the dissolving of cellular membranes (phenolics).

Sporulation: Bacteria that are resistant to environmental extremes frequently produce spores to allow survival during adverse conditions. The process of condensation of bacterial DNA within a series of membranes that accumulate calcium, dipicolinic acid and protein layers is known as sporulation. Clostidium species are known for the production of spores that allow survival of the organism under severe nutrient deprivation and dehydration conditions. Clostidium botulinum spores survive in the soil and then on harvested produce to germinate later and cause toxin-induced disease. Spores are even resistant to the actions of common antimicrobial agents and treatments.

Clostidium difficle causes disease transmission in the spore-saturated environment of hospital rooms where patients suffer from antibiotic-associated diarrhea. Species in the genus Bacillus are also spore-formers. The danger of Bacillus anthracis to the general public stems from its spore stage form and its potential for weaponization, storage and intentional release. Bacillus cereus spores germinate in warm food (especially associated with rice) and the toxins produced by the newly growing bacteria cause one form of food poisoning.

GROWTH AND REPRODUCTION

Unlike multicellular organisms, increases in the size of bacteria (cell growth) and their reproduction by cell division are tightly linked in unicellular organisms. Bacteria grow to a fixed size and then reproduce through binary fission, a form of asexual reproduction. Under optimal conditions, bacteria can grow and divide extremely rapidly and bacterial populations can double as quickly as every 9.8 minutes. In cell division, two identical clone daughter cells are produced. Some bacteria, while still reproducing asexually, form more complex reproductive structures that facilitate the dispersal of the newly-formed daughter cells. Examples include fruiting body formation by Myxobacteria and arial hyphae formation by Streptomyces, or budding. Budding involves a cell forming a protrusion that breaks away and produces a daughter cell.

In the laboratory, bacteria are usually grown using solid or liquid media. Solid growth media such as agar plates are used to isolate pure cultures of a bacterial strain. However, liquid growth media are used when measurement of growth or large volumes of cells are required. Growth in stirred liquid media occurs as an even cell suspension, making the cultures easy to divide and transfer, although isolating single bacteria from liquid media is difficult. The use of selective media (media with specific nutrients added or deficient, or with antibiotics added) can help identify specific organisms.

Most laboratory techniques for growing bacteria use high levels of nutrients to produce large amounts of cells cheaply and quickly. However, in natural environments nutrients are limited, meaning that bacteria cannot continue to reproduce indefinitely. This nutrient limitation has led the evolution of different growth strategies. Some organisms can grow extremely rapidly when nutrients become available, such as the formation of algal (and cyanobacterial) blooms that often occur in lakes during the summer.

Other organisms have adaptations to harsh environments, such as the production of multiple antibiotics by Streptomyces that inhibit the growth of competing microorganisms. In nature, many organisms live in communities (e.g. biofilms) which may allow for increased supply of nutrients and protection from environmental stresses. These relationships can be essential for growth of a particular organism or group of organisms (syntrophy).

Bacterial growth follows three phases. When a population of bacteria first enter a high-nutrient environment that allows growth, the cells need to adapt to their new environment. The first phase of growth is the lag phase, a period of slow growth when the cells are adapting to fast growth. The lag phase has high biosynthesis rates, as enzymes and nutrient transporters are produced. The second phase of growth is the logarithmic phase (log phase), also known as the exponential phase. The log phase is marked by rapid exponential growth.

The rate at which cells grow during this phase is known as the growth rate (k) and the time it takes the cells to double is known as the generation time (g). During log phase, nutrients are metabolised at maximum speed until one of the nutrients is depleted and starts limiting growth. The final phase of growth is the stationary phase and is caused by depleted nutrients. The cells reduce their metabolic activity and consume non-essential cellular proteins. The stationary phase is a transition from rapid growth to a stress response state and there is increased expression of genes involved in DNA repair, antioxidant metabolism and nutrient transport.

VIRUS REPRODUCTION

ERA in virus diseases of man was opened in 1902 by Reed and Carroll.' They and their associates recorded a number of remarkably precise observations on the etiology of yellow fever that have since been abundantly confirmed. Some of their findings bear directly on ideas which are developed in this communication. Thus, in their now classic studies with human volunteers, which proved the correctness of the theory that yellow fever was transmitted by the bite of an Aedes mosquito, they made the following observations.

Mosquitos became infective when they had taken a blood meal during the first three days of the disease. Mosquitos did not become infective when they bit early in the incubation period or late in the disease. Mosquitos were not infective soon after they had bitten and became so only after a period of 12 days or more. They then remained infective for life. These findings, so puzzling at the time, are now fully understood. Each can be explained on the basis of modem knowledge of virus reproduction and the results thereof.

Yellow fever is still one of the most classic of the virus diseases of man and it may be rewarding to review its pathogenesis briefly. When an infected mosquito bites, a very small amount of yellow fever virus is introduced into the blood. This small amount of virus then promptly disappears.

The virus does not reappear in the peripheral blood until just before the disease develops.

Therefore, normal mosquitos biting during the four-day incubation period do not take in virus. However, during the incubation period, the virus is undergoing reproduction in cells of the liver and other organs. When the number of cells damaged by the reproductive process becomes large enough, symptoms and signs of yellow fever rapidly appear. At about this time large numbers of newly formed virus particles are released from the infected tissues into the blood. Now if a mosquito bites, it will take in virus. As the disease progresses, still higher concentrations of virus develop in the liver, lymph nodes and kidneys.

This is the result of continued virus reproduction. In parallel, the concentration of virus in the blood increases rapidly and may reach a level of 107 particles per ml. If a mosquito bites at this time, it will take in a considerable amount of virus. After about the fourth day of disease, neutralizing anitibodies begin to appear and as they increase in coiicentration, the number of infective virus particles in the blood rapidly decreases and finally such particles disappear. Thus, late in the disease a biting mosquito fails to take in any infective virus and so cannot transmit the agent.

Analogous events occur in the mosquito itself. After ingestion of infective blood, the virus is first carried to the stomach of the insect and then is distributed to other organs. Only after the virus has reproduced in numerous cells in the insect does it eventually reach the salivary glands. This process takes about 12 days and during the period the mosquito, although actively supporting virus reproduction, is not mechanically able to transmit the agent. The mosquito does not develop neutralizing antibodies and, being unable to rid itself of the virus, remains infective for life. Despite extensive virus reproduction, tissue damage apparently is not produced and therefore disease does not develop in the mosquito. Although the details of pathogenesis have been documented more thoroughly for yellow fever than for any other virus disease of man, there is now sufficient information about a number of such infections to support a general conceptual scheme: only a very small number of infective virus particles are needed to initiate virus reproduction in a person who has had no previous contact with the agent.

On theoretical grounds, a single particle should, if it is adsorbed by a susceptible cell, initiate reproduction. This has, in fact, been demonstrated with vaccinia4 and influenza viruses in experimental animals. Each new virus particle from a cell can initiate reproduction in another susceptible cell. Thus, if a cell yields 100 new particles, 100 other cells of similar type can be infected and will produce a similar yield. This appears to be the basis for the logarithmic increase in virtus concentration which is the regular result of reproduction. The duration of the incubation period seems to be dependent upon two independent variables: the number of cells initially infected and the rate of virus reproduction.

The larger the number of cells initially infected, the shorter is the incubation period. The rate of virus reproduction appears to be directly related to the duration of the incubation period; the slower the rate, the longer the period. Experiments with influenza'i and other viruses in animals have demonstrated that when both variables are known the incubation period can be predicted with considerable precision. Virus disease develops only if virus reproduction occurs, but reproduction can occur without leading to disease. As a result of reproduction, the concentration of virus particles increases in infected tissue at a logarithmic rate and reaches a high level before either signs or symptoms of infection appear. This has been found to occur in every instance so far studied.

The severity of the disease may be largely determined by the concentration of virus particles in the infected tissues. Experiments have demonstrated a direct relation between virus concentration and extent of lesions with a number of viruses in animals.' From this it can be deduced that the extent of the lesions is dependent on the number of cells damaged by the reproductive process or its products. Recovery occurs when the rate of the virus reproduction decreases.

The most probable means is through reduction in the number of susceptible cells. Because recovery sometimes coincides with the appearance of antibodies against the virus, it has been held that the development of specific antibodies is responsible for recovery. There is strong evidence against this view, e.g., recovery from recurrent herpes simplex occurs without a change in antibody level, recovery from psittacosis is not associated with the appearance of significant amounts of neutralizing antibody and the injection of immune serum does not definitely lead to recovery in any virus disease.

Persistent immunity, which is of common but not uniform occurrence, is best explained by the persistence of the virus. Evidence for persistence has been obtained in a number of instances. The theory of virus persistence requires that continued reproduction occur even though this be at a very slow rate. However, it is not necessary to assume continued presence of virus in all susceptible cells to explain persistent immunity. The antigenic stimulus provided by a small amount of slowly reproducing virus would maintain antibody production and neutralizing antibodies are hiahly effective in preventing a second infection in the large majority of instances.

In the case of transient immunity, as with influenza and certain other common respiratory infections, the virus may in fact be eliminated, for these are typically surface cell infections. Although virus disease does not develop in the absence of virus reproduction, it is not a regular or even the most common sequel of the reproductive process.

Virus disease is attributable to damage of cells and the development of damage depends on the nature of the virus, its rate of reproduction and the concentration reached as well as upon the nature of the cell itself. Alteration of any one of these variables may yield a system in which reproduction occurs but disease does not develop. The easiest variable to alter is, of course, the type of cell infected. Thus, reproduction of mumps virus in the human parotid gland or meninges leads to obvious disease. Yet the same virus on comparable reproduction in the chicken embryo causes no damage."

Similarly, the reproduction of influenza viruses in the upper respiratory tract of man or the ferret often induces disease, but does not in the mouse until so-called adaptation has occurred. The effect of the rate and extent of reproduction is evident from studies with virus variants.

So-called avirulent or attenuated viruses, which do not induce obvious disease, reproduce at a slower rate and yield lower concentrations of virus than virulent strains of the same agent. Direct evidence that virus reproduction is an essential prerequisite for the development of virus disease is secured when the reproductive process is inhibited by suitable compounds. As an example, certain bacterial polysaccharides inhibit the reproduction of some small viruses in experimental animals and thereby prevent the development of disease.'

It seems clear now that the crucial problem in virus disease is, in fact, virus reproduction. If the mechanism of virus reproduction can be worked out and the process controlled, means for the management of virus diseases may be at hand. The important questions are: How do viruses multiply inside living cells and how does this process sometimes damage the cells? What are the biochemical events necessary for synthesis of new virus particles and how do they alter cell metabolism?

REPRODUCTION OF BACTERIAL VIRUSES

There is now a comprehensive scheme to explain the reproduction of these agents. It may be of value to review this in brief and then to determine how it fits in with present information on the reproduction of viruses that infect man. Bacteriophage particles resemble tiny sperm and have both a head and a tail-like structure.' The shell of the head and the tail are composed largely of protein. Inside the head there is nucleic acid and this is entirely of the desoxyribose type. A single phage particle can infect a bacterium.

Only the tail penetrates the bacterial wall and nucleic acid from the head is introduced. The phage particle itself does not enter the bacterium. In fact, the head-shell and part of the tail can be removed at this stage and yet phage reproduction will still occur in the bacterium.' This phenomenon appears to be unique in biology. The phage particle, in the process of initiating reproduction, disintegrates and its chief contribution to the bacterium is nucleic acid.

Shortly after this there are surprising developments. The bacterium stops synthesizing bacterial components and begins to produce phage nucleic acid and phage protein. The two are made separately, not as intact phage particles. After a time tailless heads appear in the bacterial cytoplasm.'

Later the final assembly of the phage particle is achieved by the addition of the tail and the introduction of nucleic acid. The new particle is now infective. The bacterial cell which supported this process is so damaged that it disintegrates and releases more than 100 new phage particles. Each new particle can initiate this complex reproductive cycle as soon as it finds another suitable uninfected bacterium.

Thus, with phage, reproduction is dependent on a series of discrete biosynthetic processes which produce the required precursor materials separately and.finally put them together as new phage particles. In supporting this remarkable process, the bacterium is driven to commit suicide.

If this concept is valid, it should be possible to control phage reproduction by inhibiting the production of any one of the required precursor materials or by preventing the final assembly of mature phage particles. Such control has, in fact, been accomplished. Chloramphenicol and analogues of tryptophane inhibit protein synthesis in bacteria. As would be expected, they prevent phage reproduction when applied during the stage of phage protein formation.

Certain pyrimidine analogues may inhibit nucleic acid synthesis in bacteria and some inhibit phage reproduction when applied early in the process. Proflavine does not prevent the synthesis of either phage protein or nucleic acid but does prevent final assembly of the mature particle. Only tailless heads are produced when this compound is applied late in the reproductive cycle and these are not infective. Apparently, however, the damage to the bacterium occurs while phage protein and nucleic acid are being formed and proflavine, although preventing assembly of the particle, does not protect the bacterial cell from destruction. It disintegrates at the expected time but does not yield infective phage particles."'

Comparison With Animal Viruses

One may now ask, in what respects does the reproduction of animal viruses correspond with the reproduction of bacterial viruses? It appears that there may be some differences, but their importance is not yet clear. Viruses which infect man resemble tiny spheres. No tail-like structures have been positively identified. They are composed largely of protein and nucleic acid, like phage particles, but some contain the ribose type, others the desoxyribose type and certain may contain both. Some have central dense areas resembling tiny nuclei." One virus particle can often initiate reproduction in a susceptible cell, but precisely how this is accomplished is not yet known. If the intact particle actually enters the cell, some change occurs, for it cannot be recovered later. Whether or not the particle actually disintegrates, as does phage, is not yet clear.

There then occurs an eclipse period during which there is no direct evidence as to intracellular developments. Some indirect evidence supports the idea that required precursor materials are formed during this period, as with phage, but their demonstration has not yet been achieved. Later, formed elements resembling virus particles begin to appear in the cytoplasm and then they increase in number. Recent evidence suggests that some viruses may develop limiting membranes before release from the cell.'

Unlike phage, the reproduction of many animal viruses does not lead to destruction of the cell. Presumably, the biochemical alterations produced in the cell are often not as severe as those caused by phage reproduction. The number of new particles produced varies from a few to more than 1,000 per cell, depending on the virus. As with phage, each new particle after release can initiate another reproductive cycle when it finds another susceptible cell. So far there are no data that make the phage scheme clearly inapplicable to the reproduction of animal viruses. If the process is like that of phage, dependent on the production of precursor materials during the eclipse period, it would be expected that reproduction should be inhibited by substances analogous to those that inhibit phage reproduction.

There is evidence in support of this idea. Thus, certain amino acid analogues, such as methoxinine or ethionine, may inhibit protein synthesis in cells. When applied early in the eclipse period, they can inhibit influenza' and poliomyelitis virus reproduction.' Moreover, some pyrimidine or purine analogues, such as diaminopurine, may inhibit nucleInsecticides

ic acid synthesis in cells. Early in the eclipse period they can inhibit the reproduction of vaccinia and encephalitis viruses.3 Acridine compounds closely related to proflavine, which prevents assembly of mature phage particles, can inhibit the reproduction of influenza and mumps virus."

Finally, certain glycosides of benzimidazole inhibit the reproduction of influenza viruses only when they are applied during the first half of the eclipse period." This is well before any new infective particles have appeared. In addition, unlike most other inhibitory compounds, some glycosides of benzimidazole do not cause demonstrable cell damage even though they are highly active as inhibitors of influenza and mumps virus reproduction.:"

Inhibition of Virus Reproduction

Thus, it appears that even as judged from the effects of chemical inhibitors of reproduction, the available evidence does not require the constrtuction of a scheme for animal virus reproduction different from that for phage reproduction. Because virus disease is dependent on virus reproduction and since reproduction can be inhibited by chemical compounds, it might be held that effective chemotherapy awaits only the finding of less toxic and more potent compounds. This idea seems sufficiently logical and so well supported by experimental evidence that an obvious difficulty may be overlooked. This difficulty was implicit even as early as 1902 in the work of Reed and Carroll' on yellow fever. It will be recalled that mosquitos became infective when they bit patients on the very first day of yellow fever. This was because a large amount of virus was already present in the blood. The virus had reached the blood only after extensive reproduction in the liver and other organs. This process occurred during the incubation period. Comparable findings have been obtainied consistently when human volunteers were inoculated with various viruses. Much virus reproduction occurs before the signs and symptoms of disease develop and on the last day of the incubation period the concentration of virus in the infected tissue is usually high, often as high as during the full-blown disease.

Experiments in animals with a variety of viruses have served merely to affirm and considerably to extend this general principle. Therefore, it appears that although inhibition of virus reproduction is a potentially feasible means for controlling virus diseases, it may need to be accomplished before the disease is fullblown if it is to be effective. To be most effective, it may prove to be necessary to inhibit virus reproduction during the incubation period.

BACTERIAL REPRODUCTION AND GENETICS

Most bacteria reproduce by a relatively simple asexual process called binary fission: each cell increases in size and divides into two cells. During this process there is an orderly increase in cellular structures and components, replication and segregation of the bacterial DNA and formation of a septum or cross wall which divides the cell into two. The process is evidently coordinated by activities associated with the cell membrane. The DNA molecule is believed to be attached to a point on the membrane where it is replicated.

The two DNA molecules remain attached at points side-by-side on the membrane while new membrane material is synthesized between the two points. This draws the DNA molecules in opposite directions while new cell wall and membrane are laid down as a septum between the two chromosomal compartments. When septum formation is complete the cell splits into two progeny cells. The time interval required for a bacterial cell to divide or for a population of cells to double is called the generation time. Generation times for bacterial species growing in nature may be as short as 15 minutes or as long as several days.

Genetic Exchange in Bacteria

Although procaryotes do not undergo sexual reproduction, they are not without the ability to exchange genes and undergo genetic recombination. Bacteria are known to exchange genes in nature by three fundamental processes: conjugation, transduction and transformation. Conjugation involves cell-to-cell contact as DNA crosses a sex pilus from donor to recipient. During transduction, a virus transfers the genes between mating bacteria. In transformation, DNA is acquired directly from the environment, having been released from another cell. Genetic recombination can follow the transfer of DNA from one cell to another leading to the emergence of a new genotype (recombinant).

It is common for DNA to be transferred as plasmids between mating bacteria. Since bacteria usually develop their genes for drug resistance on plasmids (called resistance transfer factors, or RTFs), they are able to spread drug resistance to other strains and species during genetic exchange processes. The genetic engineering of bacterial cells in the research or biotechnology laboratory is often based on the use of plasmids as vectors. The genetic systems of the Archaea are poorly characterized at this point, although the entire genome of Methanosarcina has been sequenced which opens up the possibilities for genetic analysis of the group.

EVOLUTION OF BACTERIA AND ARCHAEA

For most procaryotes, mutation is is a major source of variability that allows the species to adapt to new conditions. The mutation rate for most procaryotic genes is in the neighbourhood of 10^{-8}. This means that if a bacterial population doubles from 10^8 cells to 2×10^8 cells, there is likely to be a mutant present for any given gene. Since procaryotes grow to reach population densities far in excess of 10^9 cells, such a mutant could develop from a single generation during 15 minutes of growth. The evolution of procaryotes, driven by such Darwinian principles of evolution (mutation and selection) is called vertical evolution.

However, as a result of the processes of genetic exchange described above, the bacteria and archaea can also undergo a process of horizontal evolution. In this case, genes are transferred laterally from one organism to another, even between members of different Kingdoms, which allows the recipient to experiment with a new genetic trait. Horizontal evolution is becoming realised to be a significant force in driving cellular evolution. The combined effects of fast growth rates, high concentrations of cells, genetic processes of mutation and selection and the ability to exchange genes, account for the extraordinary rates of adaptation and evolution that can be observed in the procaryotes.

ECOLOGY OF BACTERIA AND ARCHAEA

Bacteria and Archaea are present in all environments that support life. They may be free-living, or living in associations with "higher forms" of life (plants and animals) and they are found in environments that support no other form of life. Procaryotes have the usual nutritional requirements for growth of cells, but many of the ways that they utilize and transform their nutrients are unique. This bears directly on their habitat and their ecology.

Nutritional Types of Organisms

In terms of carbon utilization a cell may be heterotrophic or autotrophic. Heterotrophs obtain their carbon and energy for growth from organic compounds in nature. Autotrophs use $C0_2$ as a sole source of carbon for growth and obtain their energy from light (e.g. photoautotrophs) or from the oxidation of inorganic compounds (e.g. lithoautotrophs).

Most heterotrophic bacteria are saprophytes, meaning that they obtain their nourishment from dead organic matter. In the soil, saprophytic bacteria and fungi are responsible for biodegradation of organic material. Ultimately, organic molecules, no matter how complex, can be degraded to CO_2 (plus H_2 and H_2O). Probably no naturally-occurring organic substance cannot be degraded by the combined activities of the bacteria and fungi. Hence, most organic matter in nature is converted by heterotrophs to CO_2, only to be converted back into organic material by autotrophs that die and nourish heterotrophs to complete the carbon cycle. Lithotrophic procaryotes have a type of energy-producing metabolism which is unique.

Lithotrophs (also called lithoautotrophs or chemoautotrophs) use inorganic compounds as sources of energy, i.e., they oxidize compounds such as H_2 or H_2S or NH_3 to obtain electrons to feed in to an electron transport system and to produce ATP. Lithotrophs are found in soil and aquatic environments wherever their energy source is present.

Most lithotrophs are autotrophs so they can grow in the absence of any organic material. Lithotrophic species are found among the Bacteria and the Archaea. Sulfur-oxidizing lithotrophs convert H_2S to S^o and S^o to SO_4. Nitrifying bacteria convert NH_3 to NO_2 and NO_2 to NO_3; methanogenic archaea strip electrons off of H_2 as a source of energy and add them to CO_2 to form CH_4 (methane). Lithotrophs have an obvious impact on the sulfur, nitrogen and carbon cycles in the biosphere. Photosynthetic bacteria convert light energy into chemical energy for growth. Most phototrophic bacteria are autotrophs so their role in the carbon cycle is analogous to that of plants. The planktonic cyanobacteria are the "grass of the sea" and their form of oxygenic photosynthesis generates a substantial amount of O_2 in the biosphere.

However, among the photosynthetic bacteria are types of photosynthetic metabolism not seen in eucaryotes, including photohetero-trophy (using light as an energy source while assimilating organic compounds as a source of carbon), anoxygenic photosynthesis and unique mechanisms of CO_2 fixation (autotrophy). Photosynthesis has not been found to occur among the Archaea, but one archaeal species employs a light-driven non photosynthetic means of energy generation based on the use of a chromophore called bacteriorhodopsin.

ENVIRONMENTAL CONDITIONS

Most procaryotes, whether they have been cultured and studied in the laboratory, or observed growing in their natural habitats, seem to be highly adapted to their specific environment by means of their macromolecular structure and/or their physiologic (metabolic) capabilities. The nutritional quality of the environment determines whether a particular organism will be present, but so do various physical parametres such as the availability of light and O_2, as well as the pH, temperature and salinity of the environment.

As examples, the range of procaryotic responses to oxygen and temperature. Procaryotes vary widely in their response to O_2 (molecular oxygen). Organisms that require O_2 for growth are called obligate aerobes; those which are inhibited or killed by O_2 and which grow only in its absence, are called obligate anaerobes; organisms which grow either in the presence or absence of O_2 are called facultative anaerobes. Whether or not a particular organism can exist in the presence of O_2 depends upon the distribution of certain enzymes such as superoxide dismutase and catalase that are required to detoxify lethal oxygen radicals that are always generated by living systems in the presence of O_2.

Procaryotes also vary widely in their response to temperature. Those that live at very cold temperatures (0 degrees or lower) are called psychrophiles; those which flourish at room temperature (25 degrees) or at the temperature of warm-blooded animals (37 degrees) are called mesophiles; those that live at high temperatures (greater than 45 degrees) are thermophiles.

The only limit that seems to be placed on growth of certain procaryotes in nature relative to temperature is whether liquid water exists. Hence, growing procaryotic cells can be found in supercooled environments (ice does not form) as low as -20 degrees and superheated environments (steam does not form) as high as 120 degrees. Archaea have been detected around thermal vents on the ocean floor where the temperature is as high as 320 degrees!

SYMBIOSIS

The biomass of procaryotic cells in the biosphere, their metabolic diversity and their persistence in all habitats that support life, ensures that these microbe will play a crucial role in the cycles of elements and the functioning of the world ecosystem. However, the procaryotes affect the world ecology in another significant way through their inevitable interactions with insects, plants and animals. Some bacteria are required to associate with insects, animals or plants for the latter to survive.

BACTERIAL PATHOGENICITY

Some bacteria are parasites of plants or animals, meaning that they grow at the expense of their eucaryotic host and may damage, harm, or even kill it in the process. Such bacteria that cause disease in plants or animals are pathogens. Human diseases caused by bacterial pathogens include tuberculosis, whooping cough, diphtheria, tetanus, gonorrhea, syphilis, pneumonia, cholera and typhoid fever, to name a few. The bacteria that cause these diseases have special structural or biochemical properties that determine their virulence or pathogenicity.

These include:

- Ability to colonize and invade their host;
- Ability to resist or withstand the antibacterial defences of the host; ability to produce various toxic substances that damage the host. Plant diseases, likewise, may be caused by bacterial pathogens. More than 200 species of bacteria are associated with plant diseases, but a very small handful of genera are involved.

Chapter 7

Lipids

FATTY ACID

One might predict that the pathway for the synthesis of fatty acids would be the reversal of the oxidation pathway. However, this would not allow distinct regulation of the two pathways to occur even given the fact that the pathways are separated within different cellular compartments.

The pathway for fatty acid synthesis occurs in the cytoplasm, whereas, oxidation occurs in the mitochondria. The other major difference is the use of nucleotide co-factors. Oxidation of fats involves the reduction of $FADH^+$ and NAD^+. Synthesis of fats involves the oxidation of NADPH. However, the essential chemistry of the two processes are reversals of each other. Both oxidation and synthesis of fats utilize an activated two carbon intermediate, acetyl-CoA. However, the acetyl-CoA in fat synthesis exists temporarily bound to the enzyme complex as malonyl-CoA.

The synthesis of malonyl-CoA is the first committed step of fatty acid synthesis and the enzyme that catalyzes this reaction, acetyl-CoA carboxylase (ACC), is the major site of regulation of fatty acid synthesis. Like other enzymes that transfer CO_2 to substrates, ACC requires a biotin co-factor.

$$\underset{\text{bicarbonate}}{HCO_3^{\ominus}} + \underset{\text{acetyl-CoA}}{H_3C{-}\overset{O}{\overset{\|}{C}}{-}SCoA} \xrightarrow[\text{ATP} \quad \text{ADP + Pi}]{\text{ACC}} \underset{\text{malonyl-CoA}}{{}^{\ominus}OOC{-}CH_2{-}\overset{O}{\overset{\|}{C}}{-}SCoA}$$

The rate of fatty acid synthesis is controlled by the equilibrium between monomeric ACC and polymeric ACC.

The activity of ACC requires polymerization. This conformational change is enhanced by citrate and inhibited by long-chain fatty acids. ACC is also controlled through hormone mediated phosphorylation. The acetyl groups that are the products of fatty acid oxidation are linked to CoASH. As you should recall, CoA contains a phosphopantetheine group coupled to AMP. The carrier of acetyl groups (and elongating acyl groups) during fatty acid synthesis is also a phosphopantetheine prosthetic group, however, it is attached a serine hydroxyl in the synthetic enzyme complex.

The carrier portion of the synthetic complex is called acyl carrier protein, ACP. This is somewhat of a misnomer in eukaryotic fatty acid synthesis since the ACP portion of the synthetic complex is simply one of many domains of a single polypeptide. The acetyl-CoA and malonyl-CoA are transferred to ACP by the action of acetyl-CoA transacylase and malonyl-CoA transacylase, respectively. The synthesis of fatty acids from acetyl-CoA and malonyl-CoA is carried out by fatty acid synthase, FAS.

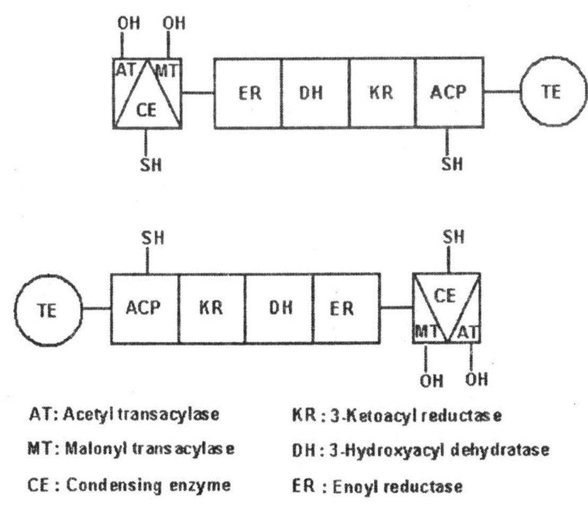

The active enzyme is a dimer of identical subunits.

All of the reactions of fatty acid synthesis are carried out by the multiple enzymatic activities of FAS. Like fat oxidation, fat synthesis involves 4 enzymatic activities. These are, b-keto-ACP synthase (condensing enzyme), b-keto-ACP reductase, 3-OH acyl-ACP dehydratase and enoyl-ACP reductase(give a look to the FAS reaction mechanism). The two reduction reactions require NADPH oxidation to $NADP^+$.

The primary fatty acid synthesized by FAS is palmitate. Palmitate is then released from the enzyme by ***thioesterase*** reaction and can then undergo separate elongation and/or unsaturation to yield other fatty acid molecules.

CYTOPLASMIC ACETYL

Acetyl-CoA is generated in the mitochondria primarily from two sources, the pyruvate dehydrogenase (PDH) reaction and fatty acid oxidation. In order for these acetyl units to be utilized for fatty acid synthesis they must be present in the cytoplasm. The shift from fatty acid oxidation and glycolytic oxidation occurs when the need for energy diminishes. This results in reduced oxidation of acetyl-CoA in the TCA cycle and the oxidative phosphorylation pathway. Under these conditions the mitochondrial acetyl units can be stored as fat for future energy demands.

Acetyl-CoA enters the cytoplasm in the form of citrate via the tricarboxylate transport system as diagrammed. In the cytoplasm, citrate is converted to oxaloacetate and acetyl-CoA by the ATP driven *ATP-citrate lyase* reaction.

This reaction is essentially the reverse of that catalyzed by the TCA enzyme *citrate synthase* except it requires the energy of ATP hydrolysis to drive it forward. The resultant oxaloacetate is converted to malate by *malate dehydrogenase* (MDH).The malate produced by this pathway can undergo oxidative decarboxylation by *malic enzyme*. The co-enzyme for this reaction is $NADP^+$ generating NADPH.

The advantage of this series of reactions for converting mitochondrial acetyl-CoA into cytoplasmic acetyl-CoA is that the NADPH produced by the *malic enzyme* reaction can be a major source of reducing co-factor for the *fatty acid synthase* activities.

REGULATION OF FATTY ACID

One must consider the global organismal energy requirements in order to effectively understand how the synthesis and degradation of fats (and also carbohydrates) needs to be exquisitely regulated. The blood is the carrier of triacylglycerols in the form of VLDLs and chylomicrons, fatty acids bound to albumin, amino acids, lactate, ketone bodies and glucose.

The pancreas is the primary organ involved in sensing the organisms dietary and energetic states via glucose concentrations in the blood. In response to low blood glucose, glucagon is secreted, whereas, in response to elevated blood glucose insulin is secreted.

The regulation of fat metabolism occurs via two distinct mechanisms. One is short term regulation which is regulation effected by events such as substrate availability, allosteric effectors and/or enzyme modification. ACC is the rate limiting (committed) step in fatty acid synthesis. This enzyme is activated by citrate and inhibited by palmitoyl-CoA and other long chain fatty acyl-CoAs. ACC activity also is affected by phosphorylation.

Glucagon stimulated increases in PKA activity result in phosphorylation of certain serine residues in ACC leading to decreased activity of the enzyme. On the other hand insulin leads to phosphorylation of ACC at sites distinct from glucagon that result in increased ACC activity. These forms of regulation are all defined as short term regulation.

Control of a given pathways' regulatory enzymes can also occur by alteration of enzyme synthesis and turn-over rates. These changes are long term regulatory effects. Insulin stimulates ACC and FAS synthesis, whereas, starvation leads to decreased synthesis of these enzymes. Adipose tissue lipoprotein lipase levels also are increased by insulin and decreased by starvation. However, in contrast to the effects of insulin and starvation on adipose tissue, their effects on heart lipoprotein lipase are just the inverse.

This allows the heart to absorb any available fatty acids in the blood in order to oxidize them for energy production.

Starvation also leads to increases in the levels of fatty acid oxidation enzymes in the heart as well as a decrease in FAS and related enzymes of synthesis.

Adipose tissue contains hormone-sensitive lipase, that is activated by PKA-dependent phosphorylation leading to increased fatty acid release to the blood. This leads to increased fatty acid oxidation in other tissues such as muscle and liver. In the liver the net result (due to increased acetyl-CoA levels) is the production of ketone bodies. This would occur under conditions where insufficient carbohydrate stores and gluconeogenic precursors were available in liver for increased glucose production.

The increased fatty acid availability in response to glucagon or epinephrine is assured of being completely oxidized since PKA also phosphorylates (and as a result inhibits) ACC, thus inhibiting fatty acid synthesis.

Insulin, on the other hand, has the opposite effect to glucagon and epinephrine leading to increased glycogen and triacylglyceride synthesis. One of the many effects of insulin is to lower cAMP levels which leads to increased dephosphorylation through the enhanced activity of protein phosphatases such as PP-1.

With respect to fatty acid metabolism this yields dephosphorylated and inactive hormone sensitive lipase. Insulin also stimulates certain phosphorylation events. This occurs through activation of several cAMP-independent kinases. Insulin stimulated phosphorylation of ACC activates this enzyme.

Regulation of fat metabolism also occurs through malonyl-CoA induced inhibition of carnitine acyltransferase I. This functions to prevent the newly synthesized fatty acids from entering the mitochondria and being oxidized.

ELONGATION AND DESATURATION

The fatty acid product released from FAS is palmitate (via the action of palmitoyl thioesterase) which is a 16:0 fatty acid, i.e. 16 carbons and no sites of unsaturation. Elongation and unsaturation of fatty acids occurs in both the mitochondria and endoplasmic reticulum (microsomal membranes).

The predominant site of these processes is in the ER membranes. Elongation involves condensation of acyl-CoA groups with malonyl-CoA.

The resultant product is two carbons longer (CO2 is released from malonyl-CoA as in the FAS reaction) which undergoes reduction, dehydration and reduction yielding a saturated fatty acid.

The reduction reactions of elongation require NADPH as co-factor just as for the similar reactions catalyzed by FAS. Mitochondrial elongation involves acetyl-CoA units and is a reversal of oxidation except that the final reduction utilizes NADPH instead of FADH2 as co-factor.

Desaturation occurs in the ER membranes as well and in mammalian cells involves 4 broad specificity fatty acyl-CoA desaturases (non-heme iron containing enzymes). These enzymes introduce unsaturation at C4, C5, C6 or C9.

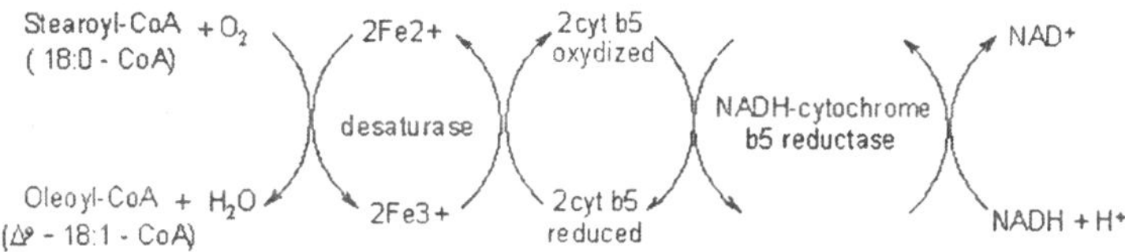

The electrons from the reduced fatty acid (saturated) and from the desaturase are transferred to molecular oxygen to form water and oxydized fatty acid (unsaturated). NADH is the donor of two electrons which keep the desaturase in the reduced form through the action of the enzyme NADH-cytochrome b5 reductase. These electrons are un-coupled from mitochondrial oxidative-phosphorylation and, therefore, do not yield ATP.

Since these enzymes cannot introduce sites of unsaturation beyond C9 they cannot synthesize either linoleate (18:2 9, 12) or linolenate (18:3 9, 12, 15). These fatty acids must be acquired from the diet and are, therefore, referred to as essential fatty acids. Linoleic is especially important in that it required for the synthesis of arachidonic acid. As we shall encounter later, arachindonate is a precursor for the eicosanoids (the prostaglandins and thromboxanes). It is this role of fatty acids in eicosanoid synthesis that leads to poor growth, wound healing and dermatitis in persons on fat free diets.

Also, linoleic acid is a constituent of epidermal cell sphingolipids that function as the skins water permeability barrier.

SYNTHESIS OF TRIGLYCERIDES

Fatty acids are stored for future use as triacylglycerols in all cells, but primarily in adipocytes of adipose tissue. Triacylglycerols constitute molecules of glycerol to which three fatty acids have been esterified.

TRIACYLGLYCEROL SYNTHESIS

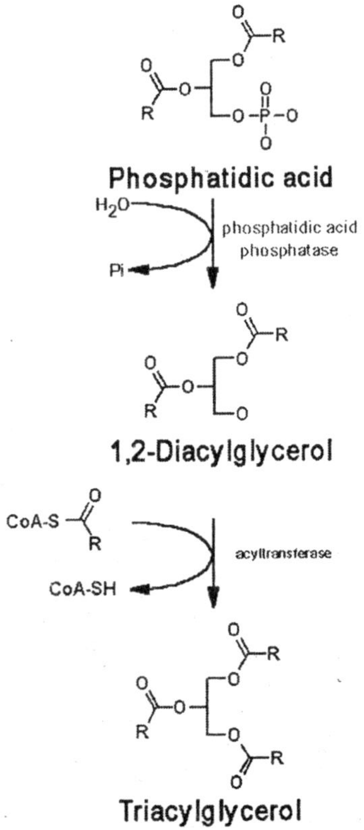

Phosphatidic acid

H_2O → Pi

phosphatidic acid phosphatase

1,2-Diacylglycerol

CoA-S–C(O)R → CoA-SH

acyltransferase

Triacylglycerol

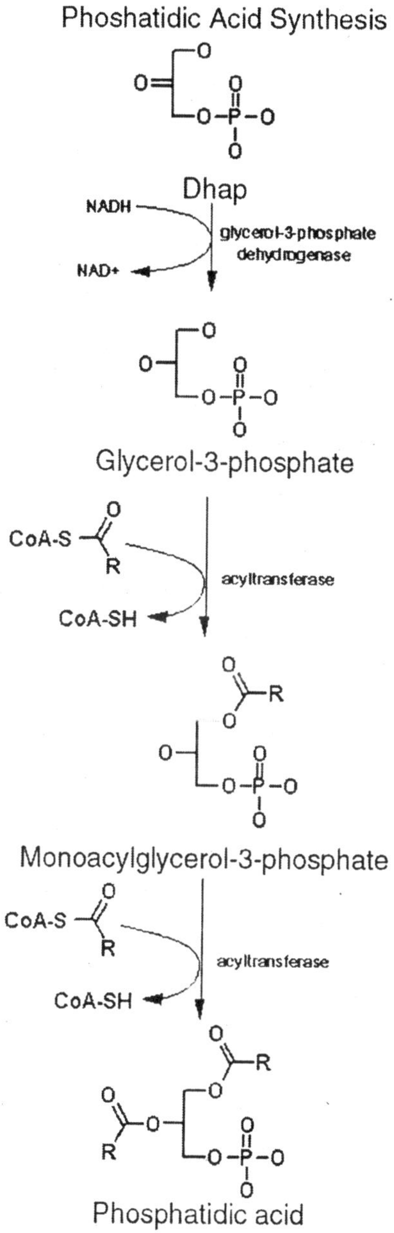

Phoshatidic Acid Synthesis

Dhap

NADH → NAD+ — glycerol-3-phosphate dehydrogenase

Glycerol-3-phosphate

CoA-S–C(=O)–R → CoA-SH — acyltransferase

Monoacylglycerol-3-phosphate

CoA-S–C(=O)–R → CoA-SH — acyltransferase

Phosphatidic acid

The fatty acids present in triacylglycerols are predominantly saturated. The major building block for the synthesis of triacylglycerols, in tissues other than adipose tissue,is glycerol-3-phosphate. Adipocytes lack glycerol kinase, therefore, dihydroxyacetone phosphate (DHAP), produced during glycolysis, is the precursor for triacylglycerol synthesis in adipose tissue.

This means that adipoctes must have glucose to oxidize in order to store fatty acids in the form of triacylglycerols. DHAP can also serve as a backbone precursor for triacylglycerol synthesis in tissues other than adipose, but does so to a much lesser extent than glycerol. The glycerol backbone of triacylglycerols is activated by phosphorylation at the C-3 position by glycerol kinase.

The utilization of DHAP for the backbone is carried out through the action of glycerol-3-phosphate dehydrogenase, a reaction that requires NADH (the same reaction as that used in the glycerol-phosphate shuttle).

The fatty acids incorporated into triacylglycerols are activated to acyl-CoAs through the action of acyl-CoA synthetases. Two molecules of acyl-CoA are esterified to glycerol-3-phosphate to yield 1,2-diacylglycerol phosphate (commonly identified as phosphatidic acid).

The phosphate is then removed, by phosphatidic acid phosphatase, to yield 1,2-diacylglycerol, the substrate for addition of the third fatty acid. Intestinal monoacylglycerols, derived from the hydrolysis of dietary fats, can also serve as substrates for the synthesis of 1,2-diacylglycerols.

PHOSPHOLIPID STRUCTURES

Phospholipids are synthesized by the addition of a basic group (predominantly a nitrogenous base) to phosphatidic acid or 1,2-diacylglycerol. Most phospholipids have a saturated fatty acid on C-1 and an unsaturated fatty acid on C-2 of the glycerol backbone. The major classifications of phospholipids are:

PHOSPHOLIPID SYNTHESIS

Phospholipids can be synthesized by two mechanisms.

One utilizes a CDP-activated polar head group for attachment to the phosphate of phosphatidic acid. The other utilizes CDP-activated 1,2-diacylglycerol and an inactivated polar head group.

PC: This class of phospholipids is also called the lecithins. At physiological pH, phosphatidylcholines are neutral zwitterions. They contain primarily palmitic or stearic acid at carbon 1 and primarily oleic, linoleic or linolenic acid at carbon 2. The lecithin dipalmitoyllecithin is a component of lung or pulmonary surfactant. It contains palmitate at both carbon 1 and 2 of glycerol and is the major (80%) phospholipid found in the extracellular lipid layer lining the pulmonary alveoli.

Choline is activated first by phosphorylation and then by coupling to CDP prior to attachment to phosphatidic acid. An other pathway to PC synthesis, involves the conversion of either PS or PE to PC. The conversion of PS to PC first requires decarboxylation of PS to yield PE; this then undergoes a series of three methylation reactions utilizing S-adenosylmethionine (SAM) as methyl group donor.

PE: These molecules are neutral zwitterions at physiological pH. They contain primarily palmitic or stearic acid on carbon 1 and a long chain unsaturated fatty acid (e.g. 18:2, 20:4 and 22:6) on carbon 2.

Synthesis of PE can occur by two pathways. The first requires that ethanolamine be activated by phosphorylation and then by coupling to CDP. The ethanolamine is then transferred from CDP-ethanolamine to phosphatidic acid to yield PE. The second involves the decarboxylation of PS.

PS: Phosphatidylserines will carry a net charge of -1 at physiological pH and are composed of fatty acids similar to the phosphatidylethanolamines.The pathway for PS synthesis involves an exchange reaction of serine for ethanolamine in PE and choline in PC. These exchange occur when PE and PC are in the lipid bilayer of the a membrane. As indicated above, PS can serve as a source of PE through a decarboxylation reaction.

PI,PG AND CARDIOLIPIN BIOSYNTHESIS

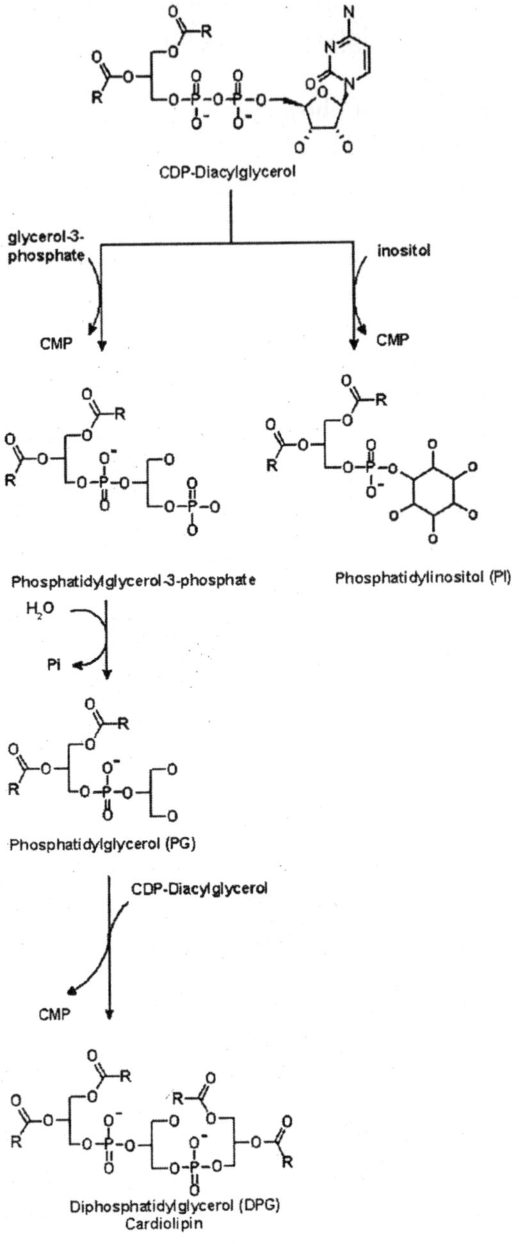

CDP-Diacylglycerol

glycerol-3-phosphate

inositol

CMP

CMP

Phosphatidylglycerol-3-phosphate

Phosphatidylinositol (PI)

H_2O

Pi

Phosphatidylglycerol (PG)

CDP-Diacylglycerol

CMP

Diphosphatidylglycerol (DPG)
Cardiolipin

PI: These molecules contain almost exclusively stearic acid at carbon 1 and arachidonic acid at carbon 2. Phosphatidylinositols composed exclusively of non-phosphorylated inositol exhibit a net charge of -1 at physiological pH. These molecules exist in membranes with various levels of phosphate esterified to the hydroxyls of the inositol. Molecules with phosphorylated inositol are termed polyphosphoinositides. The polyphosphoinositides are important intracellular transducers of signals emanating from the plasma membrane.

The synthesis of PI involves CDP-activated 1,2-diacylglycerol condensation with myo-inositol. PI subsequently undergoes a series of phosphorylations of the hydroxyls of inositol leading to the production of polyphosphoinositides. One polyphosphoinositide (phosphatidylinositol 4,5-bisphosphate, PIP2) is a critically important membrane phospholipid involved in the transmission of signals for cell growth and differentiation from outside the cell to inside.

PG: Phosphatidylglycerols exhibit a net charge of -1 at physiological pH. These molecules are found in high concentration in mitochondrial membranes and as components of pulmonary surfactant.

PG is synthesized from CDP-diacylglycerol and glycerol-3-phosphate. The vital role of PG is to serve as the precursor for the synthesis of diphosphatidylglycerols (DPGs).

DPG: These molecules are very acidic, exhibiting a net charge of -2 at physiological pH. They are found primarily in the inner mitochondrial membrane and also as components of pulmonary surfactant.DPG,also known as cardiolipins are synthesized by the condensation of CDP-diacylglycerol with PG.

The fatty acid distribution at the C-1 and C-2 positions of glycerol within phospholipids is continually in flux, owing to phospholipid degradation and the continuous phospholipid remodeling that occurs while these molecules are in membranes. Phospholipid degradation results from the action of phospholipases. There are various phospholipases that exhibit substrate specificities for different positions in phospholipids. In many cases the acyl group which was initially transferred to glycerol, by the action of the acyl transferases, is not the same acyl group present in the phospholipid when it resides within a membrane. The remodeling of acyl groups in phospholipids is the result of the action of phospholipase A_1 and phospholipase A_2.

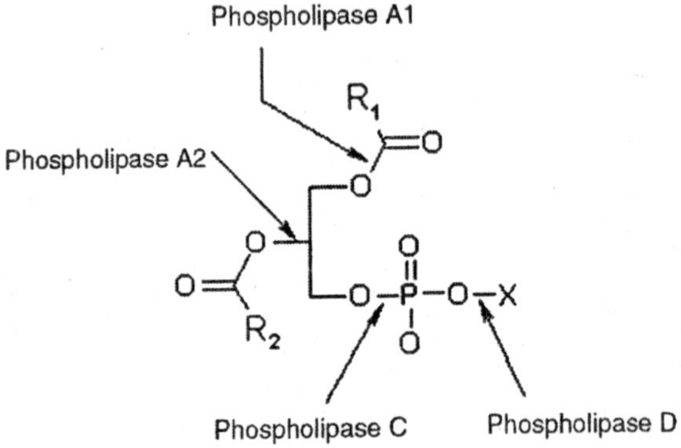

Phospholipase A1

Phospholipase A2

R_1

R_2

Phospholipase C Phospholipase D

Sites of action of the phospholipases A_1, A_2, C and D.

The products of these phospholipases are called lysophospholipids and can be substrates for acyl transferases utilizing different acyl-CoA groups. Lysophospholipids can also accept acyl groups from other phospholipids in an exchange reaction catalyzed by lysolecithin:lecithin acyltransferase (LLAT). Phospholipase A2 is also an important enzyme, whose activity is responsible for the release of arachidonic acid from the C-2 position of membrane phospholipids. The released arachidonate is then a substrate for the synthesis of the prostaglandins and leukotrienes.

METABOLISM OF SPHINGOLIPIDS

The sphingolipids, like the phospholipids, are composed of a polar head group and two nonpolar tails. The core of sphingolipids is the long-chain amino alcohol, sphingosine.

The sphingolipids include sphingomyelins and glycosphingolipids (cerebrosides, sulfatides, globosides and gangliosides). Sphingomyelins are the only sphingolipid that are also phospholipids. Sphingolipids are a component of all membranes but are particularly abundant in the myelin sheath.Ceramide,the molecule present both in sphingomyelins and glycosphingolipids derives from the condensation of serine and palmitoyl-CoA the following the pathway:

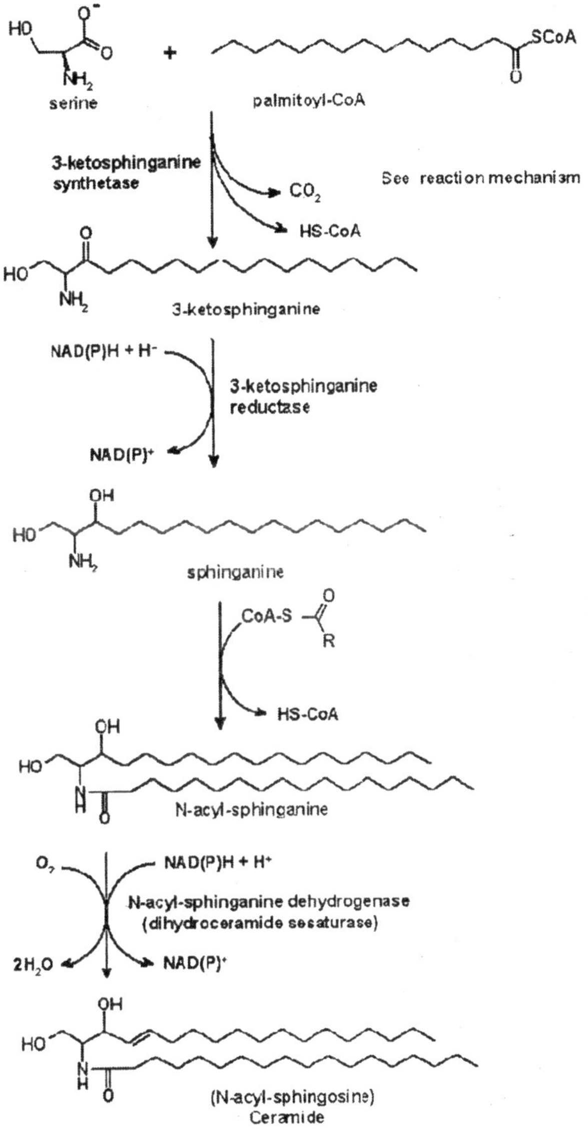

serine + palmitoyl-CoA

3-ketosphinganine synthetase

CO_2

HS-CoA

See reaction mechanism

3-ketosphinganine

NAD(P)H + H⁻

3-ketosphinganine reductase

NAD(P)⁺

sphinganine

CoA-S–C(=O)–R

HS-CoA

N-acyl-sphinganine

O_2

NAD(P)H + H⁺

N-acyl-sphinganine dehydrogenase (dihydroceramide sesaturase)

$2H_2O$

NAD(P)⁺

(N-acyl-sphingosine) Ceramide

Sphingomyelins are sphingolipids that are also phospholipids. Sphingomyelins are important structural lipid components of nerve cell membranes.

The predominant sphingomyelins contain palmitic or stearic acid N-acylated at carbon 2 of sphingosine. The sphingomyelins are synthesized by the transfer of phosphorylcholine from phosphatidylcholine to a ceramide in a reaction catalyzed by sphingomyelin synthase. Defects in the enzyme acid sphingomyelinase result in the lysosomal storage disease known as Niemann-Pick disease. There are at least 4 related disorders identified as Niemann-Pick disease Type A and B (both of which result from defects in acid sphingomyelinase), Type C1 and a related C2 and Type D. Types C1, C2 and D do not result from defects in acid sphingomyelinase.

Glycosphingolipids, or glycolipids, are composed of a ceramide backbone with a wide variety of carbohydrate groups (mono- or oligosaccharides) attached to carbon 1 of sphingosine. The four principal classes of glycosphingolipids are the cerebrosides, sulfatides, globosides and gangliosides. Cerebrosides have a single sugar group linked to ceramide. The most common of these is galactose (galactocerebrosides), with a minor level of glucose (glucocerebrosides). Galactocerebrosides are found predominantly in neuronal cell membranes. By contrast glucocerebrosides are not normally found in membranes, especially neuronal membranes; instead, they represent intermediates in the synthesis or degradation of more complex glycosphingolipids. Galactocerebrosides are synthesized from ceramide and UDP-galactose. Excess accumulation of glucocerebrosides is observed in Gaucher's disease.

Sulfatides: The sulfuric acid esters of galactocerebrosides are the sulfatides. Sulfatides are synthesized from galactocerebrosides and activated sulfate, 3'-phosphoadenosine 5'-phosphosulfate (PAPS). Excess accumulation of sulfatides is observed in sulfatide lipidosis.

Globosides: Globosides represent cerebrosides that contain additional carbohydrates, predominantly galactose, glucose or GalNAc. Lactosyl ceramide is a globoside found in erythrocyte plasma membranes. Globotriaosylceramide (also called ceramide trihexoside) contains glucose and two moles of galactose and accumulates, primarily in the kidneys, of patients suffering from Fabry's disease.

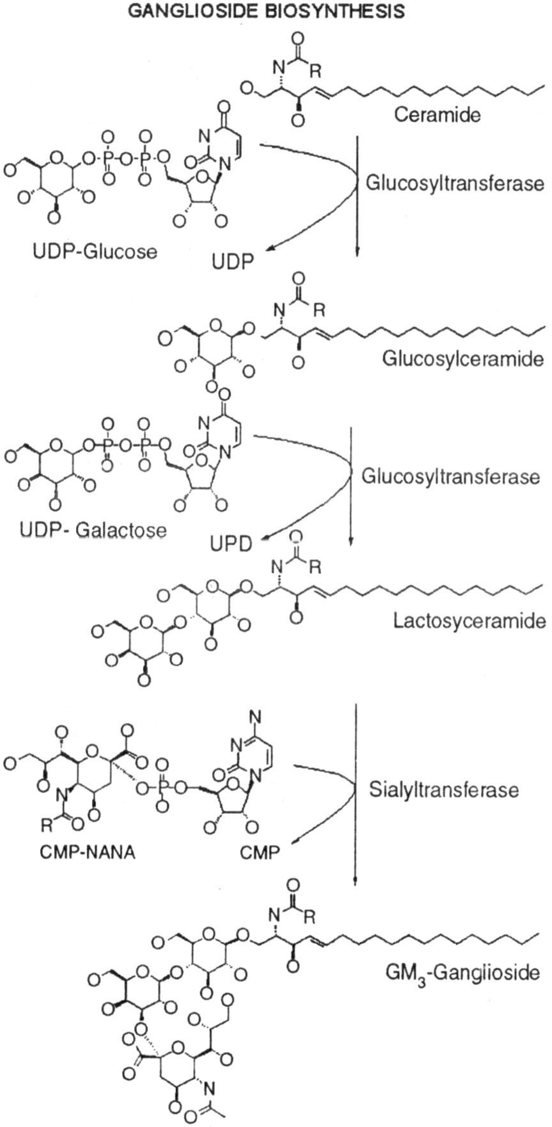

GANGLIOSIDE BIOSYNTHESIS

Ceramide

UDP-Glucose UDP

Glucosyltransferase

Glucosylceramide

UDP- Galactose UPD

Glucosyltransferase

Lactosyceramide

CMP-NANA CMP

Sialyltransferase

GM_3-Ganglioside

Gangliosides: Gangliosides are very similar to globosides except that they also contain NANA in varying amounts. The specific names for gangliosides are a key to their structure. The letter G refers to"ganglioside", and the subscripts M, D, T and Q indicate that the molecule contains mono-, di-, tri and quatra(tetra)-sialic acid.

The numerical subscripts 1, 2 and 3 refer to the carbohydrate sequence that is attached to ceramide;1 stands for Gal-GalNAc-Gal-Glc-ceramide, 2 for GalNAc-Gal-Glc-ceramide and 3 for Gal-Glc-ceramide. Deficiencies in lysosomal enzymes, which normally are responsible for the degradation of the carbohydrate portions of various gangliosides, underlie the symptoms observed in rare autosomally inherited diseases termed lipid storage diseases, many of which are listed below.

CLINICAL SIGNIFICANCES OF SPHINGOLIPIDS

One of the most clinically important classes of sphingolipids are those that confer antigenic determinants on the surfaces of cells, particularly the erythrocytes. The ABO blood group antigens are the carbohydrate moieties of glycolipids on the surface of cells as well as the carbohydrate portion of serum glycoproteins. When present on the surface of cells the ABO carbohydrates are linked to sphingolipid and are therefore of the glycosphingolipid class. When the ABO carbohydrates are associated with protein in the form of glycoproteins they are found in the serum and are referred to as the seThe eicosanoids consist of thecreted forms.

Some individuals produce the glycoprotein forms of the ABO antigens while others do not. This property distinguishes secretors from non-secretors, a property that has forensic importance such as in cases of rape. *Strucures of the ABO blood group carbohydrates:* A significant cause of death in premature infants and, on occasion, in full term infants is respiratory distress syndrome (RDS) or hyaline membrane disease. This condition is caused by an insufficient amount of pulmonary surfactant. Under normal conditions the surfactant is synthesized by type II endothelial cells and is secreted into the alveolar spaces to prevent atelectasis following expiration during breathing.Surfactant is comprised primarily of dipalmitoyllecithin; additional lipid components include phosphatidylglycerol and phosphatidylinositol along with proteins of 18 and 36 kDa (termed surfactant proteins). During the third trimester the fetal lung synthesizes primarily sphingomyelin, and type II endothelial cells convert the majority of their stored glycogen to fatty acids and then to dipalmitoyllecithin.

Fetal lung maturity can be determined by measuring the ratio of lecithin to sphingomyelin (L/S ratio) in the amniotic fluid. An L/S ratio less than 2.0 indicates a potential risk of RDS. The risk is nearly 75-80% when the L/S ratio is 1.5.

The carbohydrate portion of the ganglioside, GM 1, present on the surface of intestinal epithelial cells, is the site of attachment of cholera toxin, the protein secreted by Vibrio cholerae.

These are just a few examples of how sphingolipids and glycosphingolipids are involved in various recognition functions at the surface of cells. As with the complex glycoproteins, an understanding of all of the functions of the glycolipids is far from complete.

ABNORMAL SPHINGOLIPID METABOLISM

Based upon genetic linkage analyses as well as enzyme studies and the characterization of accumulating lysosomal substances, Niemann Pick disease should be divided into type I and type II; type I has 2 subtypes, A and B (257200), which show deficiency of acid sphingomyelinase. Niemann Pick disease type II likewise has 2 subtypes, type C1 and C2 (NPC) and type D (NPD). It is obviously confusing to use the abbreviation NPD for Niemann Pick disease in some cases and for subtype D of Niemann Pick disease in other cases.

This gene contains regions of homology to mediators of cholesterol homeostasis suggesting why LDL-cholesterol accumulates in lysosomes of afflicted individuals. The encoded protein product of NPC1 gene is 1278 amino acids long.

Disorder	Enzyme Deficiency	Accumulating Substance	Symptoms
Tay-Sachs disease	Hexosaminidase A	GM_2 ganglioside	mental retardation, blindness, early mortality
Gaucher's disease	Glucocerebrosidase	Glucocerebroside	hepatosplenomegaly, mental retardation in infantile form, long bone degeneration
Fabry's disease	α-Galactosidase A	Globotriaosylceramide; also called ceramide trihexoside (CTH)	kidney failure, skin rashes
Niemann-Pick disease, more info below Types A and B Type C1 Type C2 Type D	Sphingomyelinase see info below see info below	Sphingomyelin LDL-derived cholesterol LDL-derived cholesterol	all types lead to mental retardation, hepatosplenomegaly, early fatality potential
Krabbe's disease; globoid leukodystrophy	Galactocerebrosidase	Galactocerebroside	mental retardation, myelin deficiency
Sandhoff-Jatzkewitz disease	Hexosaminidase A and B	Globoside, GM_2 ganglioside	same symptoms as Tay-Sachs,progresses more rapidly
GM_1 gangliosidosis	GM_1 ganglioside: β - galactosidase	GM_1 ganglioside	mental retardation, skeletal abnormalities, hepatomegaly
Sulfatide lipodosis; metachromatic leukodystrophy	Arylsulfatase A	Sulfatide	mental retardation, metachromasia of nerves
Fucosidosis	α-L-Fucosidase	Pentahexosylfucoglycolipid	cerebral degeneration, thickened skin, muscle spasticity
Farber's lipogranulomatosis	Acid ceramidase	Ceramide	hepatosplenomegaly, painful swollen joints

Within the protein are regions of homology to the

transmembrane domain of the morphogen receptor PATCHED (of Drosophila melanogaster), the putative sterol-sensing regions of SREBP (sterol regulatory element binding protein) cleavage-activating protein (SCAP) and HMG-CoA reductase.

METABOLISM OF THE EICOSANOIDS

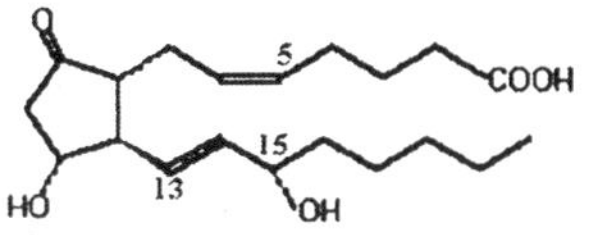

PGE_2

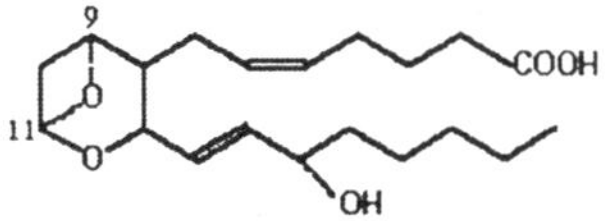

TXA_2

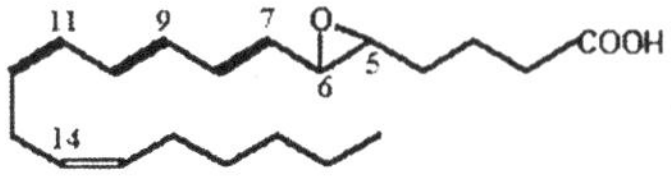

LTA_4

The eicosanoids consist of the prostaglandins (PGs), thromboxanes (TXs) and leukotrienes (LTs). The PGs and TXs are collectively identified as prostanoids. Prostaglandins were originally shown to be synthesized in the prostate gland, thromboxanes from platelets (thrombocytes) and leukotrienes from leukocytes, hence the derivation of their names.

The eicosanoids produce a wide range of biological effects on inflammatory responses (predominantly those of the joints, skin and eyes), on the intensity and duration of pain and fever, and on reproductive function (including the induction of labour).

They also play important roles in inhibiting gastric acid secretion, regulating blood pressure through vasodilation or constriction, and inhibiting or activating platelet aggregation and thrombosis.

The principal eicosanoids of biological significance to humans are a group of molecules derived from the C20 fatty acid, arachidonic acid. Minor eicosanoids are derived from dihomo-g-linoleic acid and eicosopentaenoic acid. The major source of arachidonic acid is through its release from cellular stores.

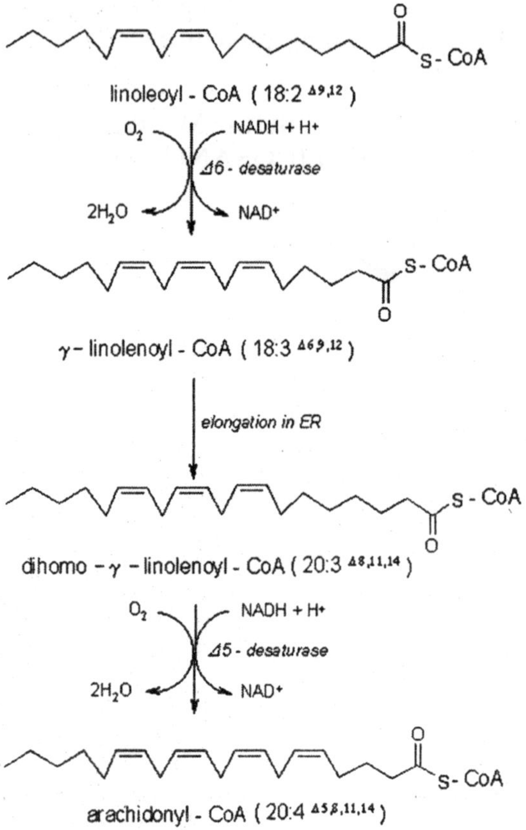

S-CoA

linoleoyl - CoA (18:2 $^{\Delta 9,12}$)

O_2 NADH + H+

Δ6 - desaturase

$2H_2O$ NAD^+

S-CoA

γ- linolenoyl - CoA (18:3 $^{\Delta 6,9,12}$)

elongation in ER

S - CoA

dihomo - γ - linolenoyl - CoA (20:3 $^{\Delta 8,11,14}$)

O_2 NADH + H+

Δ5 - desaturase

$2H_2O$ NAD^+

S- CoA

arachidonyl - CoA (20:4 $^{\Delta 5,8,11,14}$)

Within the cell, it resides predominantly at the C-2 position of membrane phospholipids and is released from there upon the activation of phospholipase A2 . The immediate dietary precursor of arachidonate is linoleate. Linoleate is also the precursor for dihomo-g-linoleic acid and eicosopentaenoic acid synthesis. Therefore, the absence of linoleic acid from the diet would seriously threaten the body's ability to synthesize eicosanoids.

Pathway from linoleic acid to arachidonic acid. Numbers in parentheses refer to the fatty acid length and the number and positions of unsaturations.

All mammalian cells except erythrocytes synthesize eicosanoids. These molecules are extremely potent, able to cause profound physiological effects at very dilute concentrations. All eicosanoids function locally at the site of synthesis, through receptor-mediated G-protein linked signaling pathways leading to an increase in cAMP levels.

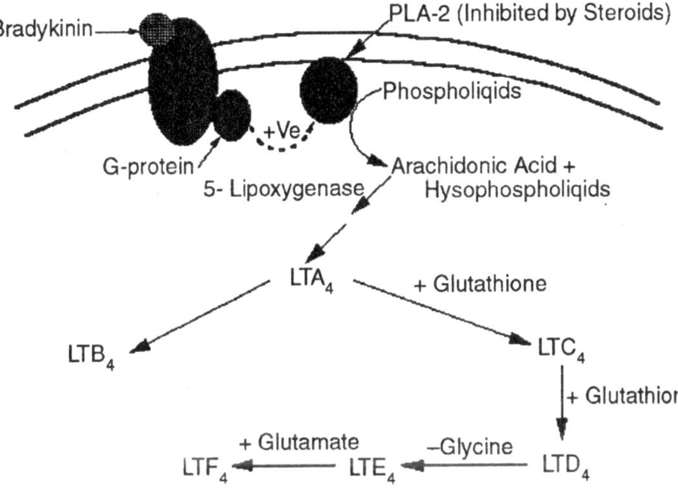

Two main pathways are involved in the biosynthesis of eicosanoids. The prostaglandins and thromboxanes are synthesized by the cyclic pathway, the leukotrienes by the linear pathway. Synthesis of the clinically relevant prostaglandins and thromboxanes from arachidonic acid. Numerous stimuli (e.g. epinephrine, thrombin and bradykinin) activate phospholipase A2 which hydrolyzes arachidonic acid from membrane phospholipids. The prostaglandins are identified as PG and the thromboxanes as TX. Prostaglandin PGI2 is also known as prostacyclin. The subscript 2 in each molecule refers to the number of -C=C- present. The cyclic pathway is initiated through the action of prostaglandin endoperoxide synthetase. This enzyme possesses two activities, cyclooxygenase and peroxidase. The linear pathway is initiated through the action of lipoxygenases. It is the enzyme, 5-lipoxygenase that gives rise to the leukotrienes.

A widely used class of drugs, the non-steroidal anti-inflammatory drugs (NSAIDs such as ibuprofen, indomethacin, naproxen, phenylbutazone and aspirin, all act upon the cyclooxygenase activity. Another class, the corticosteroidal drugs, act to inhibit phospholipase A2, thereby inhibiting the release of arachidonate from membrane phospholipids and the subsequent synthesis of eicosinoids.

FATTY ACID OXIDATION

Utilization of dietary lipids requires that they first be absorbed through the intestine. As these molecules are oils they would be essentially insoluble in the aqueous intestinal environment. Solubilization (emulsification) of dietary lipid is accomplished via bile salts that are synthesized in the liver and secreted from the gallbladder.

The emulsified fats can then be degraded by pancreatic lipases (lipase and phospholipase A2). These enzymes, secreted into the intestine from the pancreas, generate free fatty acids and a mixtures of mono- and diacylglycerols from dietary triacylglycerols. Pancreatic lipase degrades triacylglycerols at the 1 and 3 positions sequentially to generate 1,2-diacylglycerols and 2-acylglycerols. Phospholipids are degraded at the 2 position by pancreatic phospholipase A2 releasing a free fatty acid and the lysophospholipid.

Following absorption of the products of pancreatic lipase by the intestinal mucosal cells, the resynthesis of triacylglycerols occurs. The triacylglycerols are then solubilized in lipoprotein complexes (complexes of lipid and protein) called chylomicrons. A chylomicron contains lipid droplets surrounded by the more polar lipids and finally a layer of proteins.

Triacylglycerols synthesized in the liver are packaged into VLDLs and released into the blood directly. Chylomicrons from the intestine are then released into the blood via the lymph system for delivery to the various tissues for storage or production of energy through oxidation.

The triacylglycerol components of VLDLs and chylomicrons are hydrolyzed to free fatty acids and glycerol in the capillaries of adipose tissue and skeletal muscle by the action of lipoprotein lipase. The free fatty acids are then absorbed by the cells and the glycerol is returned via the blood to the liver (and kidneys). The glycerol is then converted to the glycolytic intermediate DHAP.

The classification of blood lipids is distinguished based upon the density of the different lipoproteins. As lipid is less dense than protein, the lower the density of lipoprotein the less protein there is.

MOBILIZATION OF FAT STORES

The primary sources of fatty acids for oxidation are dietary and mobilization from cellular stores. Fatty acids from the diet can are delivered from the gut to cells via transport in the blood.

Fatty acids are stored in the form of triacylglycerols primarily within adipocytes of adipose tissue. In response to energy demands, the fatty acids of stored triacylglycerols can be mobilized for use by peripheral tissues. The release of metabolic energy, in the form of fatty acids, is controlled by a complex series of interrelated cascades that result in the activation of hormone-sensitive lipase.

The stimulus to activate this cascade, in adipocytes, can be glucagon, epinephrine or b-corticotropin. These hormones bind cell-surface receptors that are coupled to the activation of adenylate cyclase upon ligand binding. The resultant increase in cAMP leads to activation of PKA, which in turn phosphorylates and activates hormone-sensitive lipase. This enzyme hydrolyzes fatty acids from carbon atoms 1 or 3 of triacylglycerols. The resulting diacylglycerols are substrates for either hormone-sensitive lipase or for the non-inducible enzyme diacylglycerol lipase. Finally the monoacylglycerols are substrates for monoacylglycerol lipase. The net result of the action of these enzymes is three moles of free fatty acid and one mole of glycerol.

The free fatty acids diffuse from adipose cells, combine with albumin in the blood, and are thereby transported to other tissues, where they passively diffuse into cells.

Hormone Induced Fatty Acid Mobilization in Adipocytes

Model for the activation of hormone-sensitive lipase by epinephrine. Epinephrine binds its' receptor and leads to the activation of adenylate cyclase.

The resultant increase in cAMP activates PKA which then phosphorylates and activates hormone-sensitive lipase. Hormone-sensitive lipase hydrolyzes fatty acids from triacylglycerols and diacylglycerols.

The final fatty acid is released from monoacylglycerols through the action of monoacylglycerol lipase, an enzyme active in the absence of hormonal stimulation.

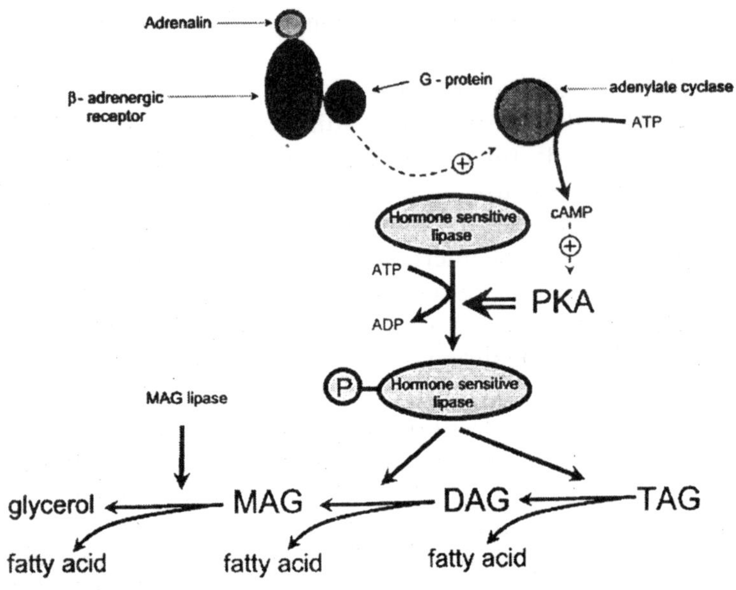

In contrast to the hormonal activation of adenylate cyclase and (subsequently) hormone-sensitive lipase in adipocytes, the mobilization of fat from adipose tissue is inhibited by numerous stimuli.

The most significant inhibition is that exerted upon adenylate cyclase by insulin. When an individual is well fed state, insulin released from the pancreas prevents the inappropriate mobilization of stored fat. Instead, any excess fat and carbohydrate are incorporated into the triacylglycerol pool within adipose tissue.

ALTERNATIVE OXIDATION PATHWAYS

The majority of natural lipids contain an even number of carbon atoms. A small proportion that contain odd numbers; upon complete -oxidation, these yield acetyl-CoA units plus a single mole of propionyl-CoA. The propionyl-CoA is converted, in an ATP-dependent pathway, to succinyl-CoA. The succinyl-CoA can then enter the TCA cycle for further oxidation.

The oxidation of unsaturated fatty acids is essentially the same process as for saturated fats, except when a double bond is encountered. In such a case, the bond is isomerized by a specific enoyl-CoA isomerase and oxidation continues. In the case of linoleate, the presence of the -12 unsaturation results in the formation of a dienoyl-CoA during oxidation. This molecule is the substrate for an additional oxidizing enzyme, the NADPH requiring 2,4-dienoyl-CoA reductase.

Phytanic acid is a fatty acid present in the tissues of ruminants and in dairy products and is, therefore, an important dietary component of fatty acid intake. Because phytanic acid is methylated, it cannot act as a substrate for the first enzyme of the -oxidation pathway (acyl-CoA dehydrogenase). An additional mitochondrial enzyme, -hydroxylase, adds a hydroxyl group to the -carbon of phytanic acid, which then serves as a substrate for the remainder of the normal oxidative enzymes. This process is termed -oxidation.

REGULATION OF FATTY ACID METABOLISM

In order to understand how the synthesis and degradation of fats needs to be exquisitely regulated, one must consider the energy requirements of the organism as a whole. The blood is the carrier of triacylglycerols in the form of VLDLs and chylomicrons, fatty acids bound to albumin, amino acids, lactate, ketone bodies and glucose. The pancreas is the primary organ involved in sensing the organism's dietary and energetic states by monitoring glucose concentrations in the blood. Low blood glucose stimulates the secretion of glucagon, whereas, elevated blood glucose calls for the secretion of insulin.

The metabolism of fat is regulated by two distinct mechanisms. One is short-term regulation, which can come about through events such as substrate availability, allosteric effectors and/or enzyme modification. The other mechanism, long-term regulation, is achieved by alteration of the rate of enzyme synthesis and turn-over.

ACC (acetyl-CoA carboxylase) is the rate-limiting (committed) step in fatty acid synthesis. This enzyme is activated by citrate and inhibited by palmitoyl-CoA and other long-chain fatty acyl-CoAs. ACC activity can also be affected by phosphorylation. For instance, glucagon-stimulated increases in PKA activity result in the phosphorylation of certain serine residues in ACC leading to decreased activity of the enzyme. By contrast, insulin leads to PKA-independent phosphorylation of ACC at sites distinct from glucagon, which bring about increased ACC activity. Both of these reaction chains are examples of short-term regulation.

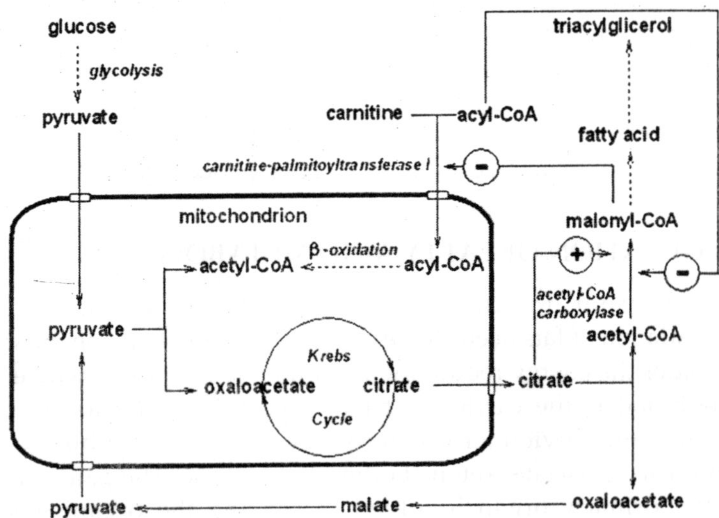

Insulin, a product of the well-fed state, stimulates ACC and FAS synthesis, whereas starvation leads to a decrease in the synthesis of these enzymes. Adipose tissue levels of lipoprotein lipase also are increased by insulin and decreased by starvation. However, the effects of insulin and starvation on lipoprotein lipase in the heart are just the inverse of those in adipose tissue.

This sensitivity allows the heart to absorb any available fatty acids in the blood in order to oxidize them for energy production. Starvation also leads to increases in the levels of cardiac enzymes of fatty acid oxidation, and to decreases in FAS and related enzymes of synthesis.

Adipose tissue contains hormone-sensitive lipase, which is activated by PKA-dependent phosphorylation; this activation increases the release of fatty acids into the blood. This in turn leads to the increased oxidation of fatty acids in other tissues such as muscle and liver. In the liver, the net result (due to increased acetyl-CoA levels) is the production of ketone bodies. This would occur under conditions in which the carbohydrate stores and gluconeogenic precursors available in the liver are not sufficient to allow increased glucose production. The increased levels of fatty acid that become available in response to glucagon or epinephrine are assured of being completely oxidized, because PKA also phosphorylates ACC; the synthesis of fatty acid is thereby inhibited.

Insulin has the opposite effect to glucagon and epinephrine: it increases the synthesis of triacylglycerols (and glycogen). One of the many effects of insulin is to lower cAMP levels, which leads to increased dephosphorylation through the enhanced activity of protein phosphatases such as PP-1. With respect to fatty acid metabolism, this yields dephosphorylated and inactive hormone-sensitive lipase. Insulin also stimulates certain phosphorylation events. This occurs through activation of several cAMP-independent kinases, one of which phosphorylates and thereby stimulates the activity of ACC.

Fat metabolism can also be regulated by malonyl-CoA-mediated inhibition of carnitine acyltransferase I. Such regulation serves to prevent de novo synthesized fatty acids from entering the mitochondria and being oxidized.

CLINICAL SIGNIFICANCE OF FATTY ACIDS

The majority of clinical problems related to fatty acid metabolism are associated with processes of oxidation.

These disorders fall into four main groups:

- *Deficiencies in Carnitine*: Deficiencies in carnitine lead to an inability to transport fatty acids into the mitochondria for oxidation. This can occur in newborns and particularly in pre-term infants. Carnitine deficiencies also are found in patients undergoing hemodialysis or exhibiting organic aciduria. Carnitine deficiencies may manifest systemic symptomology or may be limited to only muscles. Symptoms can range from mild occasional muscle cramping to severe weakness or even death. Treatment is by oral carnitine administration.
- *Carnitine Palmitoyltransferase I (CPT I) Deficiency*: Deficiencies in this enzyme affect primarily the liver and lead to reduced fatty acid oxidation and ketogenesis. Carnitine Palmitoylransferase II (CPT II) deficiency results in recurrent muscle pain and fatigue and myoglobinuria following strenuous exercise. Carnitine acyltransferases may also be inhibited by sulfonylurea drugs such as tolbutamide and glyburide.
- *Deficiencies in Acyl-CoA Dehydrogenases:* A group of inherited diseases that impair -oxidation result from deficiencies in acyl-CoA dehydrogenases.

The enzymes affected may belong to one of four categories:

- Very long-chain acyl-CoA dehydrogenase (VLCAD)
- Long-chain acyl-CoA dehydrogenase (LCAD)
- Medium-chain acyl-CoA dehydrogenase (MCAD)
- Short-chain acyl-CoA dehydrogenase (SCAD)

MCAD deficiency is the most common form of this disease. In the first years of life this deficiency will become apparent following a prolonged fasting period. Symptoms include vomiting, lethargy and frequently coma. Excessive urinary excretion of medium-chain dicarboxylic acids as well as their glycine and carnitine esters is diagnostic of this condition. In the case of this enzyme deficiency. taking care to avoid prolonged fasting is sufficient to prevent clinical problems.

- *Refsum's Disease*: Refsum's disease is a rare inherited disorder in which patients lack the mitochondrial - oxidizing enzyme. As a consequence, they accumulate large quantities of phytanic acid in their tissues and serum. This leads to severe symptoms, including cerebellar ataxia, retinitis pigmentosa, nerve deafness and peripheral neuropathy. As expected, the restriction of dairy products and ruminant meat from the diet can ameliorate the symptoms of this disease.

KETOGENESIS

During high rates of fatty acid oxidation, primarily in the liver, large amounts of acetyl-CoA are generated. These exceed the capacity of the TCA cycle, and one result is the synthesis of ketone bodies, or ketogenesis. The ketone bodies are acetoacetate, -hydroxybutyrate, and acetone.

The formation of acetoacetyl-CoA occurs by condensation of two moles of acetyl-CoA through a reversal of the thiolase catalyzed reaction of fat oxidation. Acetoacetyl-CoA and an additional acetyl-CoA are converted to -hydroxy- -methylglutaryl-CoA (HMG-CoA) by HMG-CoA synthase, an enzyme found in large amounts only in the liver. Some of the HMG-CoA leaves the mitochondria, where it is converted to mevalonate (the precursor for cholesterol synthesis) by HMG-CoA reductase. HMG-CoA in the mitochondria is converted to acetoacetate by the action of HMG-CoA lyase. Acetoacetate can undergo spontaneous decarboxylation to acetone, or be enzymatically converted to -hydroxybutyrate through the action of -hydroxybutyrate dehydrogenase. When the level of glycogen in the liver is high the production of -hydroxybutyrate increases.

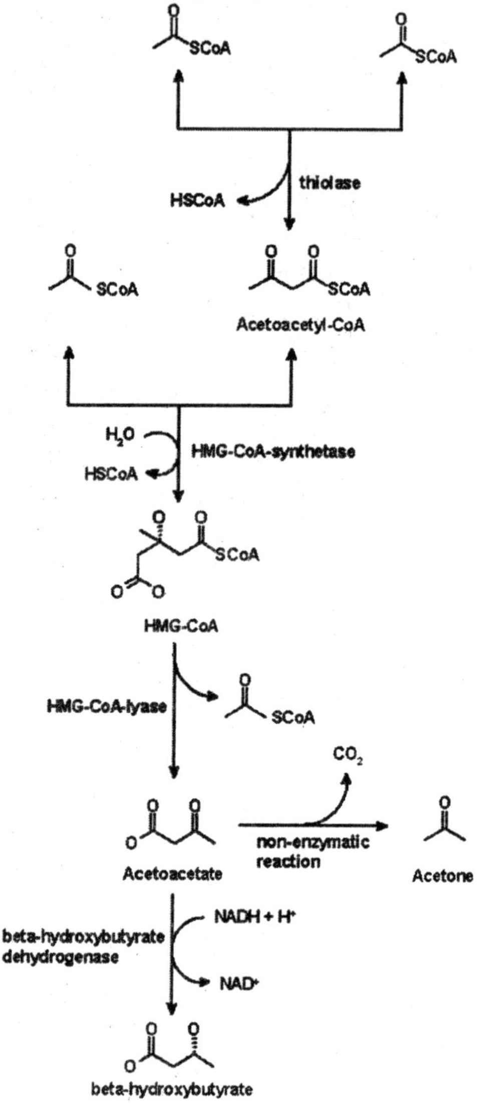

When carbohydrate utilization is low or deficient, the level of oxaloacetate will also be low, resulting in a reduced flux through the TCA cycle. This in turn leads to increased release of ketone bodies from the liver for use as fuel by other tissues. In early stages of starvation, when the last remnants of fat are oxidized, heart and skeletal muscle will consume primarily ketone bodies to preserve glucose for use by the brain. Acetoacetate and -hydroxybutyrate, in particular, also serve as major substrates for the biosynthesis of neonatal cerebral lipids.

Ketone bodies are utilized by extrahepatic tissues through the conversion of -hydroxybutyrate to acetoacetate and of acetoacetate to acetoacetyl-CoA. The first step involves the reversal of the -hydroxybutyrate dehydrogenase reaction, and the second involves the action (shown below) of acetoacetate:succinyl-CoA transferase, also called ketoacyl-CoA-transferase.

Acetoacetate + Succinyl-CoA ↔ Acetoacetyl-CoA + succinate

The latter enzyme is present in all tissues except the liver. Importantly, its absence allows the liver to produce ketone bodies but not to utilize them. This ensures that extrahepatic tissues have access to ketone bodies as a fuel source during prolonged starvation. Acetoacetyl-CoA is converted into two molecules of acetyl-CoA by a thiolase:

Acetoacetyl-CoA + HS-CoA ↔ 2 Acetyl-CoA

REGULATION OF KETOGENESIS

The fate of the products of fatty acid metabolism is determined by an individual's physiological status. Ketogenesis takes place primarily in the liver and may by affected by several factors:

- Control in the release of free fatty acids from adipose tissue directly affects the level of ketogenesis in the liver. This is, of course, substrate-level regulation.
- Once fats enter the liver, they have two distinct fates. They may be activated to acyl-CoAs and oxidized, or esterified to glycerol in the production of triacylglycerols. If the liver has sufficient supplies of glycerol-3-phosphate, most of the fats will be turned to the production of triacylglycerols.

- The generation of acetyl-CoA by oxidation of fats can be completely oxidized in the TCA cycle. Therefore, if the demand for ATP is high the fate of acetyl-CoA is likely to be further oxidation to CO_2.
- The level of fat oxidation is regulated hormonally through phosphorylation of ACC, which may activate it (in response to glucagon) or inhibit it (in the case of insulin).

CLINICAL SIGNIFICANCE OF KETOGENESIS

The production of ketone bodies occurs at a relatively low rate during normal feeding and under conditions of normal physiological status.

Normal physiological responses to carbohydrate shortages cause the liver to increase the production of ketone bodies from the acetyl-CoA generated from fatty acid oxidation. This allows the heart and skeletal muscles primarily to use ketone bodies for energy, thereby preserving the limited glucose for use by the brain. The most significant disruption in the level of ketosis, leading to profound clinical manifestations, occurs in untreated insulin-dependent diabetes mellitus. This physiological state, diabetic ketoacidosis (DKA), results from a reduced supply of glucose (due to a significant decline in circulating insulin) and a concomitant increase in fatty acid oxidation (due to a concomitant increase in circulating glucagon).

The increased production of acetyl-CoA leads to ketone body production that exceeds the ability of peripheral tissues to oxidize them. Ketone bodies are relatively strong acids (pKa around 3.5), and their increase lowers the pH of the blood. This acidification of the blood is dangerous chiefly because it impairs the ability of hemoglobin to bind oxygen.

CHOLESTEROL AND BILE METABOLISM

CHOLESTEROL METABOLISM

Cholesterol is an extremely important biological molecule that has roles in membrane structure as well as being a precursor for the synthesis of the steroid hormones and bile acids. Both dietary cholesterol and that synthesized de novo are transported through the circulation in lipoprotein particles. The same is true of cholesteryl esters, the form in which cholesterol is stored in cells. The synthesis and utilization of cholesterol must be tightly regulated in order to prevent over-accumulation and abnormal deposition within the body. Of particular importance clinically is the abnormal deposition of cholesterol and cholesterol-rich lipoproteins in the coronary arteries. Such deposition, eventually leading to atherosclerosis, is the leading contributory factor in diseases of the coronary arteries

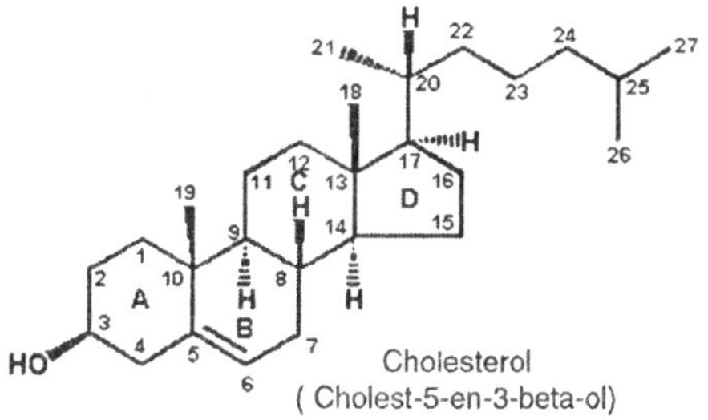

Cholesterol
(Cholest-5-en-3-beta-ol)

BIOSYNTHESIS OF CHOLESTEROL

Slightly less than half of the cholesterol in the body derives from biosynthesis de novo.

Biosynthesis in the liver accounts for approximately 10%, and in the intestines approximately 15%, of the amount produced each day. Cholesterol synthesis occurs in the cytoplasm and microsomes from the two-carbon acetate group of acetyl-CoA.

The process has nine major steps:

- Acetyl-CoAs are converted to ω3-hydroxy- 3-methyl glutaryl-CoA (HMG-CoA).Unlike the HMG-CoA formed during ketone body synthesis in the mitochondria, this form is synthesized in the cytoplasm. However, the pathway and the necessary enzymes are the same as those in the mitochondria.
- Two moles of acetyl-CoA are condensed in a reversal of the thiolase reaction, forming acetoacetyl-CoA.

Acetoacetyl-CoA and a third mole of acetyl-CoA are converted to HMG-CoA by the action of HMG-CoA synthase.

- HMG-CoA is converted to mevalonate by HMG-CoA reductase (this enzyme is bound to the endoplasmic reticulum). HMG-CoA reductase absolutely requires NADPH as a cofactor and two moles of NADPH are consumed during the conversion of HMG-CoA to mevalonate.
- The reaction catalyzed by HMG-CoA reductase is the rate limiting step of cholesterol biosynthesis, and this enzyme is subject to complex regulatory controls.

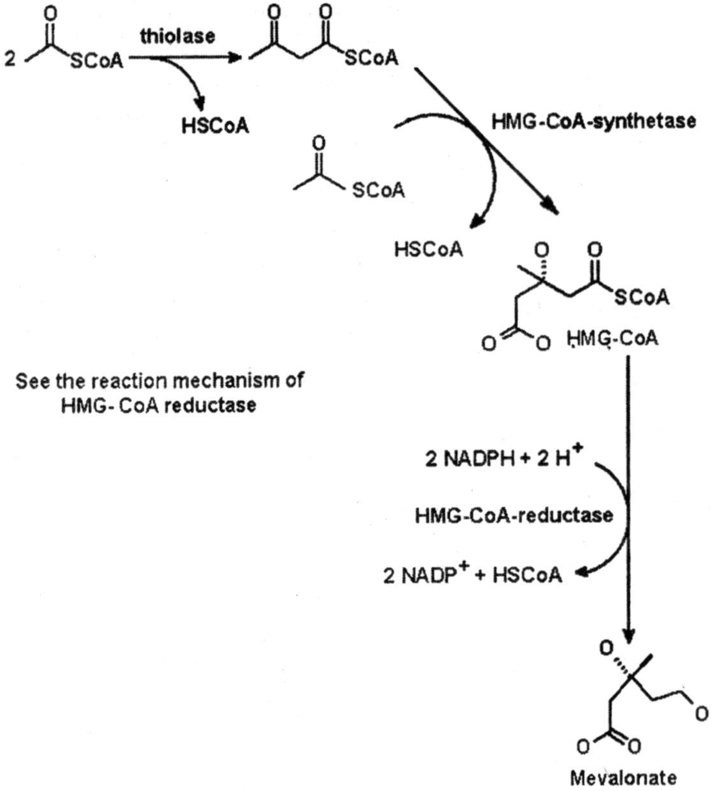

2 SCoA → (thiolase) SCoA + HSCoA

SCoA → (HMG-CoA-synthetase) HSCoA

SCoA

HMG-CoA

See the reaction mechanism of HMG- CoA reductase

2 NADPH + 2 H^+

HMG-CoA-reductase

2 $NADP^+$ + HSCoA

Mevalonate

- Mevalonate is then activated by two successive phosphorylations, yielding 5-pyrophosphomevalonate.

- In addition to activating mevalonate, the phosphorylations maintain its solubility, since otherwise it is insoluble in water.
- GPP further condenses with another IPP molecule to yield farnesyl pyrophosphate, (FPP).

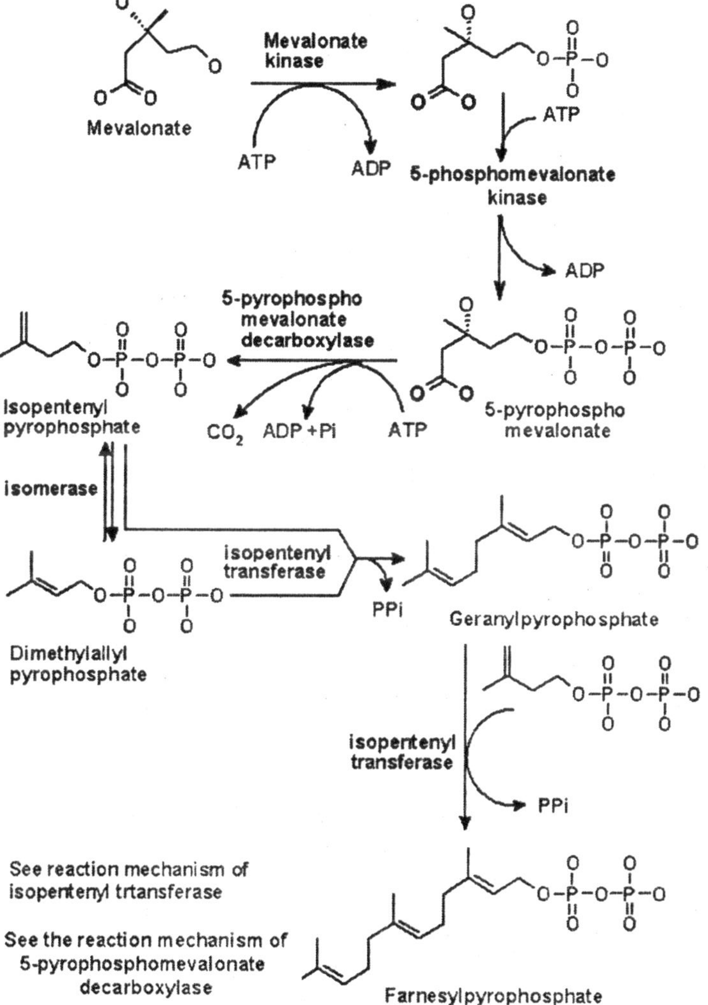

- Mevalonate is converted to the isoprene based molecule, isopentenyl pyrophosphate (IPP), with the concomitant loss of CO_2. Isopentenyl pyrophosphate is in equilibrium with its isomer, dimethylallyl pyrophosphate, (DMPP).

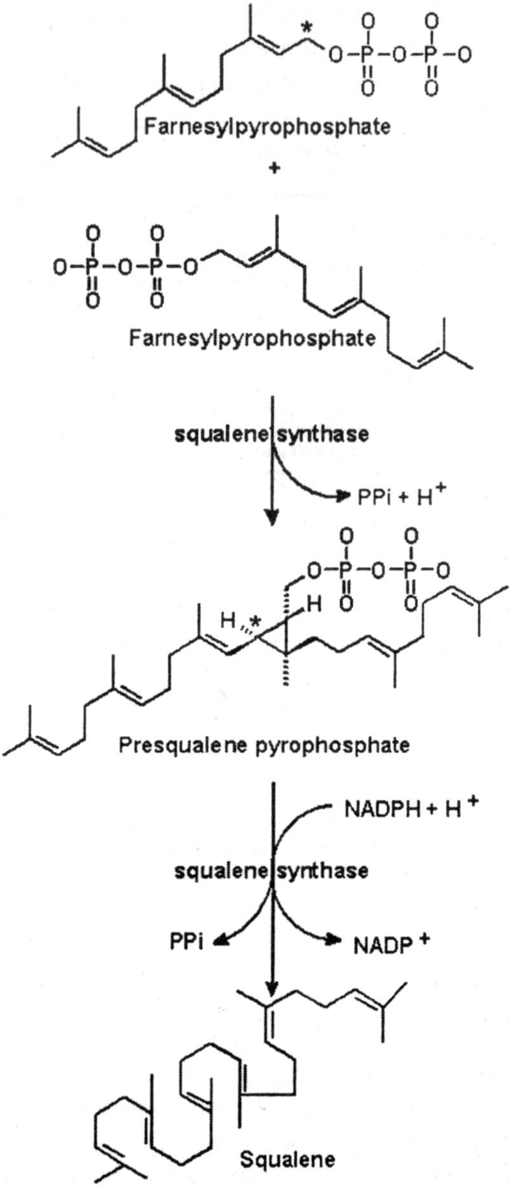

Farnesylpyrophosphate

+

Farnesylpyrophosphate

squalene synthase

PPi + H^+

Presqualene pyrophosphate

$NADPH + H^+$

squalene synthase

PPi

$NADP^+$

Squalene

- One molecule of IPP condenses with one molecule of DMPP to generate geranyl pyrophosphate, (GPP).

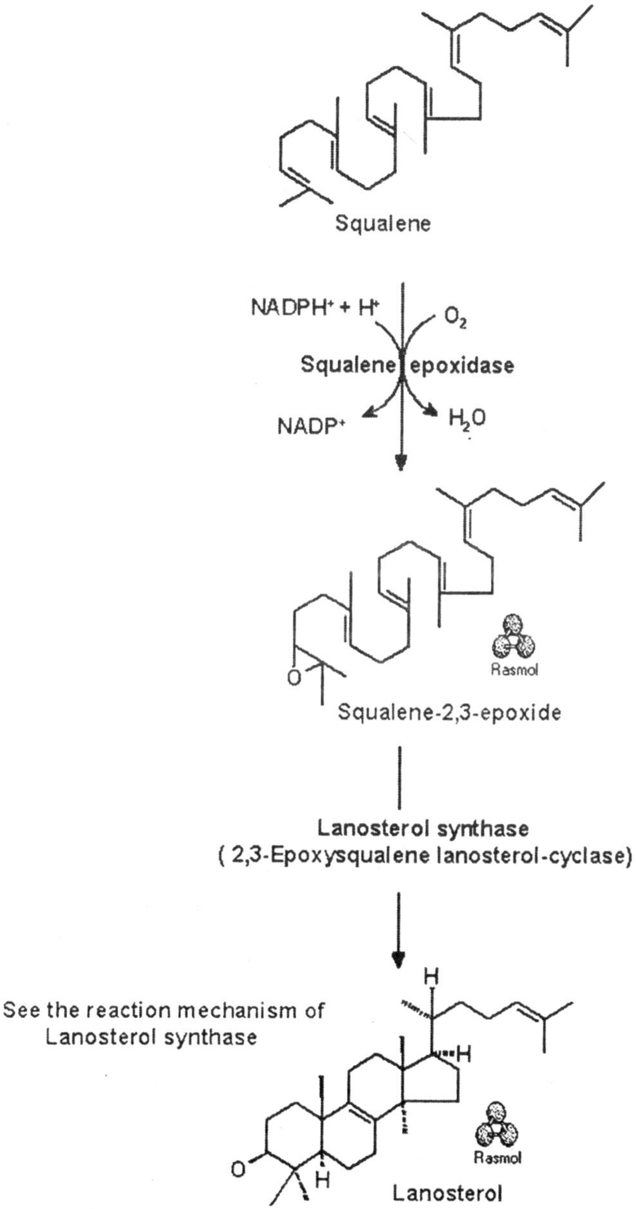

Squalene

$NADPH^+ + H^+$ O_2

Squalene epoxidase

$NADP^+$ H_2O

Rasmol

Squalene-2,3-epoxide

Lanosterol synthase
(2,3-Epoxysqualene lanosterol-cyclase)

See the reaction mechanism of Lanosterol synthase

H H O H

Rasmol

Lanosterol

- Finally, the NADPH-requiring enzyme, squalene synthase catalyzes the head-to-tail condensation of two molecules of FPP, yielding squalene (squalene synthase also is tightly associated with the endoplasmic reticulum).
- Squalene undergoes a two step cyclization to yield lanosterol. The first reaction is catalyzed by squalene monooxygenase. This enzyme uses NADPH as a cofactor to introduce molecular oxygen as an epoxide at the 2,3 position of squalene.
- Through a series of 19 additional reactions, lanosterol is converted to cholesterol.

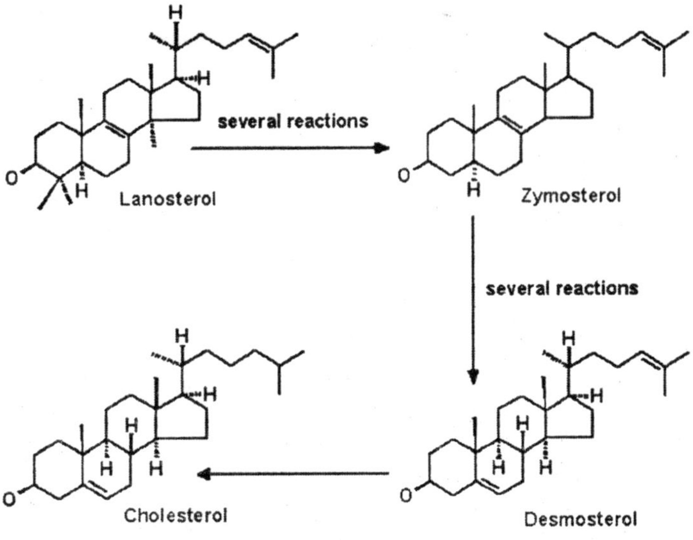

Lanosterol → several reactions → Zymosterol → several reactions → Desmosterol → Cholesterol

PATHWAY OF CHOLESTEROL BIOSYNTHESIS

Synthesis begins with the transport of acetyl-CoA from the mitochondrion to the cytosol. The rate limiting step occurs at the 3-hydroxy-3-methylglutaryl-CoA (HMG-CoA) reducatse catalyzed step. The phosphorylation reactions are required to solubilize the isoprenoid intermediates in the pathway. Intermediates in the pathway are used for the synthesis of prenylated proteins, dolichol, coenzyme Q and the side chain of heme a.

The acetyl-CoA utilized for cholesterol biosynthesis is derived from an oxidation reaction (eg, fatty acids or pyruvate) in the mitochondria and is transported to the cytoplasm by the same process as that described for fatty acid synthesis. Acetyl-CoA can also be derived from cytoplasmic oxidation of ethanol by acetyl-CoA synthetase.

All the reduction reactions of cholesterol biosynthesis use NADPH as a cofactor. The isoprenoid intermediates of cholesterol biosynthesis can be diverted to other synthesis reactions, such as those for dolichol (used in the synthesis of N-linked glycoproteins, coenzyme Q (of the oxidative phosphorylation) pathway or the side chain of heme a. Additionally, these intermediates are used in the lipid modification of some proteins.

Regulating Cholesterol Synthesis

Normal healthy adults synthesize cholesterol at a rate of approximately 1g/day and consume approximately 0.3g/day. A relatively constant level of cholesterol in the body (150 - 200 mg/dL) is maintained primarily by controlling the level of de novo synthesis. The level of cholesterol synthesis is regulated in part by the dietary intake of cholesterol. Cholesterol from both diet and synthesis is utilized in the formation of membranes and in the synthesis of the steroid hormones and bile acids. The greatest proportion of cholesterol is used in bile acid synthesis.

The cellular supply of cholesterol is maintained at a steady level by three distinct mechanisms:

- Regulation of HMG-CoA reductase activity and levels
- Regulation of excess intracellular free cholesterol through the activity of acyl-CoA:cholesterol acyltransferase, ACAT
- Regulation of plasma cholesterol levels via LDL receptor-mediated uptake and HDL-mediated reverse transport.

Regulation of HMG-CoA reductase activity is the primary means for controlling the level of cholesterol biosynthesis.

Regulation of HMG-CoA reductase by covalent

modification. HMG-CoA reductase is most active in the dephosphorylated state. Phosphorylation is catalyzed by reductase kinase (RK), an enzymes whose activity is also regulated by phosphorylation.

Phosphorylation of RK is catalyzed by reductase kinase kinase (RKK). Hormones such as glucagon and epinephrine negatively affect cholesterol biosynthesis by increasing the activity of the inhibitor of phosphoprotein phosphatase-1, PPI-1.

Conversely, insulin stimulates the removal of phosphates and, thereby, activates HMG-CoA reductase activity. Additional regulation of HMG-CoA reductase occurs through an inhibition of synthesis of the enzyme by elevation in intracellular cholesterol levels.

Regulation of HMG-CoA reductase takes place both through covalent modification and through control of the absolute level of the enzyme within cells. HMG-CoA reductase is a single polypeptide embedded in microsomal membranes; it is most active in its unmodified form.

Phosphorylation of the enzyme decreases its activity.HMG-CoA reductase is phosphorylated by reductase kinase, RK. RK itself is activated via phosphorylation. The phosphorylation of RK is catalyzed by reductase kinase kinase, RKK. There are two isoforms of RKK, one independent of cAMP and one dependent upon cAMP. The cAMP-dependent RKK is activated in the presence of cAMP. Since the intracellular level of cAMP is regulated by hormonal stimuli, regulation of cholesterol biosynthesis is hormonally controlled.

Insulin leads to a decrease in cAMP, which in turn activates cholesterol synthesis. Alternatively, glucagon and epinephrine--- which increase the level of cAMP--- inhibit cholesterol synthesis. The activity of HMG-CoA reductase is further controlled by the cAMP signaling pathway. Increases in cAMP lead to phosphorylation and activation of phosphoprotein phosphatase inhibitor-1 (PPI-1). HMG-CoA reductase is dephosphorylated and, thereby, activated by phosphoprotein phosphatase-1 (PP-1).

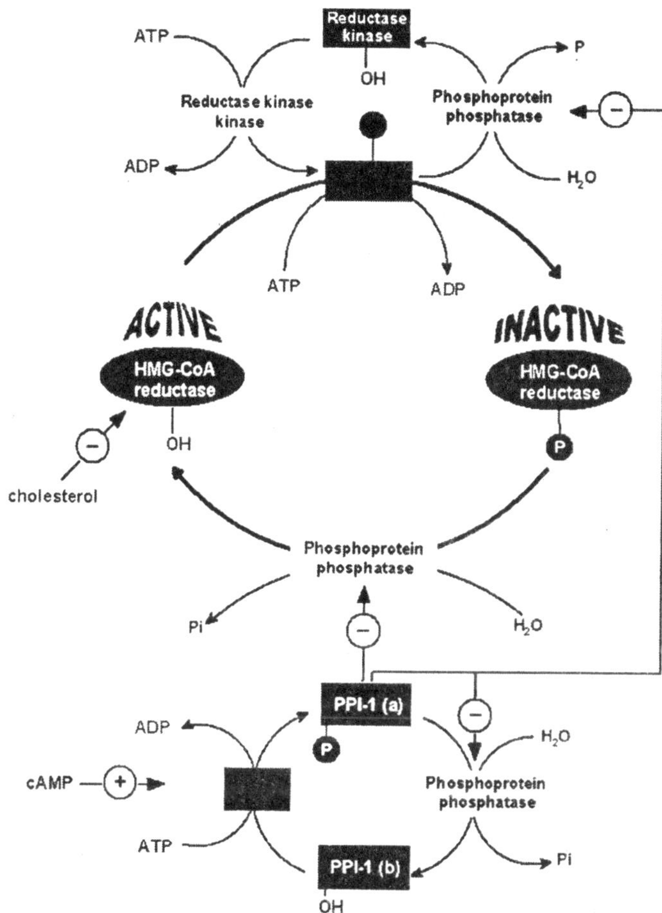

At the same time, PP-1 dephosphorylates and inactivates RK. The cAMP-induced activation of PPI-1 results in a decrease in the level of active PP-1, resulting in a reduced ability of PP-1 to dephosphorylate HMG-CoA reductase and RK. This maintains RK in the phosphorylated and active state, and HMG-CoA reductase in the phosphorylated and inactive state. As the stimulus leading to increased cAMP production is removed, the level of phosphorylations decreases and that of dephosphorylations increases. The net result is a return to a higher level of HMG-CoA reductase activity.

This regulatory cascade is additionally affected by PKA-mediated phosphorylation of PPI, which leads to reduced dephosphorylation by PP-1.The ability of insulin to stimulate, and glucagon to inhibit, HMG-CoA reductase activity is consistent with the effects of these hormones on other metabolic pathways.

The basic function of these two hormones is to control the availability and delivery of energy to all cells of the body.Long-term control of HMG-CoA reductase activity is exerted primarily through control over the synthesis and degradation of the enzyme. When levels of cholesterol are high, the level of expression of the HMG-CoA gene is reduced. Conversely, reduced levels of cholesterol activate expression of the gene. Insulin also brings about long-term regulation of cholesterol metabolism by increasing the level of HMG-CoA reductase synthesis. The rate of HMG-CoA turn-over is also regulated by the supply of cholesterol. When cholesterol is abundant, the rate of HMG-CoA reductase degradation increases.

REGULATION OF HMG-COA REDUCTASE

The stability of HMGR is regulated as the rate of flux through the mevalonate synthesis pathway changes. When the flux is high the rate of HMGR degradation is also high. When the flux is low, degradation of HMGR decreases. This phenomenon can easily be observed in the presence of the statin drugs. HMGR is localized to the ER and like SREBP contains a sterol-sensing domain,SSD. When sterol levels increase in cells there is an concommitant increase in the rate of HMGR degradation. The degradation of HMGR occurs within the proteosome, a multiprotein complex dedicated to protein degradation. The primary signal directing proteins to the proteosome is ubiquitination.

Ubiquitin is a 7.6kDa protein that is covalently attached to proteins targeted for degradation by ubiqitin ligases. These enzymes attach multiple copies of ubiquitin allowing for recognition by the proteosome. HMGR has been shown to be ubiquitinated prior to its degradation. The primary sterol regulating HMGR degradation is cholesterol itself. As the levels of free cholesterol increase in cells, the rate of HMGR degradation increases.

The Utilization of Cholesterol

Cholesterol is transported in the plasma predominantly as cholesteryl esters associated with lipoproteins. Dietary cholesterol is transported from the small intestine to the liver within chylomicrons. Cholesterol synthesized by the liver, as well as any dietary cholesterol in the liver that exceeds hepatic needs, is transported in the serum within LDLs. The liver synthesizes VLDLs and these are converted to LDLs through the action of endothelial cell-associated lipoprotein lipase.

Cholesterol found in plasma membranes can be extracted by HDLs and esterified by the HDL-associated enzyme LCAT. The cholesterol acquired from peripheral tissues by HDLs can then be transferred to VLDLs and LDLs via the action of cholesteryl ester transfer protein (apo-D) which is associated with HDLs. Reverse cholesterol transport allows peripheral cholesterol to be returned to the liver in LDLs. Ultimately, cholesterol is excreted in the bile as free cholesterol or as bile salts following conversion to bile acids in the liver.

Regulation of Cellular Sterol Content

The continual alteration of the intracellular sterol content occurs through the regulation of key sterol synthetic enzymes as well as by altering the levels of cell-surface LDL receptors. As cells need more sterol they will induce their synthesis and uptake, conversely when the need declines synthesis and uptake are decreased.

Regulation of these events is brought about primarily by sterol-regulated transcription of key rate limiting enzymes and by the regulated degradation of HMGR. Activation of transcriptional control occurs through the regulated cleavage of the membrane-bound transcription factor sterol regulated element binding protein, SREBP. As discussed above, degradation of HMGR is controlled by the ubiquitin-mediated pathway for proteolysis.

Sterol control of transcription affects more than 30 genes involved in the biosynthesis of cholesterol, triacylglycerols, phospholipids and fatty acids. Transcriptional control requires the presence of an octamer sequence in the gene termed the sterol regulatory element, SRE-1.

It has been shown that SREBP is the transcription factor that binds to SRE-1 elements. It turns out that there are 2 distinct SREBP gene, SREBP-1 and SREBP-2. In addition, the SREBP-1 gene encodes 2 proteins, SREBP-1a and SREBP-1c as a consequence of alternative exon usage. All 3 proteins are proteolytically regulated by sterols. Full-length SREBPs have several domains and are embedded in the membrane of the endoplasmic reticulum (ER).

The N-terminal domain contains a transcription factor motif of the basic helix-loop-helix (bHLH) type that is exposed to the cytoplasmic side of the ER. There are 2 transmembrane spanning domains followed by a large C-terminal domain also exposed to the cytosolic side.

When sterols are scarce, cleavage of the full-length SREBP takes place with the result being that the N-terminla bHLH motif is released into the cytosol. The bHLH domian then migrates to the nucleus to direct transcription. Conversely, when sterols are abundant, cleavage of SREBP is inhibited. To control the level of SREBP-mediated transcription, the soluble bHLH domain is itself subject to rapid proteolysis.The cleavage of SREBP is carried out by 2 distinct enzymes, one of which is regulated by sterols.

The regulated cleavage occurs in the lumenal loop between the 2 transmembrane domains. This cleavage is catalyzed by site-1 protease, S1P. High sterol content blocks the activity of S1P. The second cleavage, catalyzed by site-2 protease, S2P, occurs in the first transmembrane span, leading to release of active SREBP. In order for S2P to act on SREBP site-1 must already have been cleaved.

Additional studies on sterol-regulated gene expression demonstrated that cleavage of SREBP by S1P is controlled by the level and action of an additional protein termed, SREBP cleavage-activating protein, SCAP. SCAP is a large protein also found in the ER membrane and contains at least 8 transmembrane spans. The C-terminal portion, which extends into the cytosol, has been shown to interact with the C-terminal domain of SREBP. This C-terminal region of SCAP contains 4 motifs called WD40 repeats. The WD40 repeats are required for interaction of SCAP with SREBP.

Interestingly, the N-terminus of SCAP, including membrane spans 2-6, resembles HMGR which itself is subject to sterol-stimulated degradation.

This shared motif is called the sterol sensing domain, SSD. Several proteins whose functions involve streols also contain the SSD. These include patched, an important development regulating receptor whose ligand, hedgehog, is modified by attachment of cholesterol and the Niemann Pick C1 (NPC1) protein which is involved in cholesterol transport in the secretory pathway.

The function of SCAP is to positively stimulate S1P-mediated cleavage of SREBP. The function of sterols is to inhibit this positive action of SCAP. The activity of SCAP involves movement from the ER to the Golgi and back. Because the C-terminus of SCAP interacts with SREBP, movement of SCAP takes SREBP along for the ride. When sterols are low, SCAP and SREBP move to the Golgi. This transit is required for SREBP cleavage as S1P is Golgi-localized. When sterols are high, movement of SCAP is halted. Thus, the overall effect of sterols is to regulate the ability of SCAP to present SREBP to S1P.

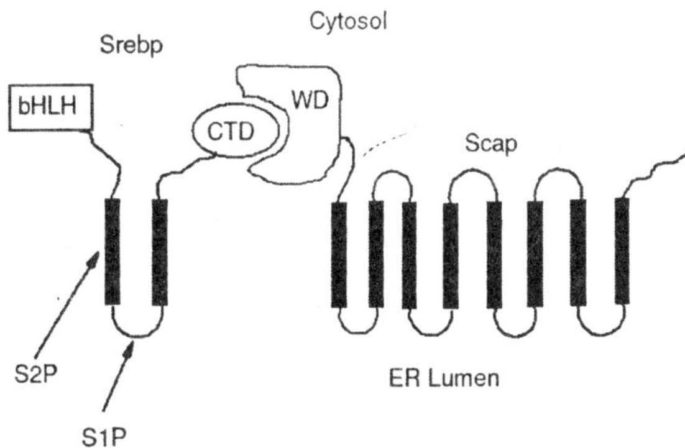

Diagramatic representation of the interactions between SREBP and SCAP in the membrane of the ER.

CTD = C-terminal Domain.

BILE ACIDS SYNTHESIS AND UTILIZATION

The end products of cholesterol utilization are the bile acids, synthesized in the liver. Synthesis of bile acids is the predominant mechanisms for the excretion of excess cholesterol.

However, the excretion of cholesterol in the form of bile acids is insufficient to compensate for an excess dietary intake of cholesterol.

Synthesis of the 2 primary bile acids, cholic acid and chenodeoxycholic acid. The reaction catalyzed by the 7a-hydroxylase is the rate limiting step in bile acid synthesis. Conversion of 7 -hydroxycholesterol to the bile acids requires several steps not shown in detail in this image.

The most abundant bile acids in human bile are chenodeoxycholic acid (45%) and cholic acid (31%). These are referred to as the primary bile acids. Within the intestines the primary bile acids are acted upon by bacteria and converted to the secondary bile acids, identified as deoxycholate (from cholate) and lithocholate (from chenodeoxycholate). Both primary and secondary bile acids are reabsorbed by the intestines and delivered back to the liver via the portal circulation.

Within the liver the carboxyl group of primary and secondary bile acids is conjugated via an amide bond to either glycine or taurine before their being re-secreted into the bile canaliculi.

These conjugation reactions yield glycoconjugates and tauroconjugates, respectively. The bile canaliculi join with the bile ductules, which then form the bile ducts. Bile acids are carried from the liver through these ducts to the gallbladder, where they are stored for future use.

The ultimate fate of bile acids is secretion into the intestine, where they aid in the emulsification of dietary lipids. In the gut the glycine and taurine residues are removed and the bile acids are either excreted (only a small per centage) or reabsorbed by the gut and returned to the liver. This process of secretion from the liver to the gallbladder, to the intestines and finally re-absorbtion is termed the enterohepatic circulation.

Cholesterol

↓

7α-hydroxycholesterol

↓

7α-hydroxycholest-4-en-3-one

Several steps

Choloyl-CoA

Chenodeoxycholoyl-CoA

CLINICAL SIGNIFICANCE OF BILE ACID SYNTHESIS

Bile acids perform four physiologically significant functions:

- Their synthesis and subsequent excretion in the feces represent the only significant mechanism for the elimination of excess cholesterol.
- Bile acids and phospholipids solubilize cholesterol in the bile, thereby preventing the precipitation of cholesterol in the gallbladder.
- They facilitate the digestion of dietary triacylglycerols by acting as emulsifying agents that render fats accessible to pancreatic lipases.
- They facilitate the intestinal absorption of fat-soluble vitamins.

MASTER METABOLIC REGULATOR

AMP-activated protein kinase (AMPK) was first discovered as an activity that inhibited preparations of acetyl-CoA carboxylase (ACC) and 3-hydroxy-3-methylglutaryl-CoA reductase (HMG-CoA reductase, HMGR) and was induced by AMP. AMPK induces a cascade of events within cells in response to the ever changing energy charge of the cell. The role of AMPK in regulating cellular energy charge places this enzyme at a central control point in maintaining energy homeostasis. More recent evidence has shown that AMPK activity can also be regulated by physiological stimuli, independent of the energy charge of the cell, including hormones and nutrients.

Once activated, AMPK-mediated phosphorylation events switch cells from active ATP consumption (e.g. fatty acid and cholesterol biosynthesis) to active ATP production (e.g. fatty acid and glucose oxidation). These events are rapidly initiated and are referred to as short-term regulatory processes. The activation of AMPK also exerts long-term effects at the level of both gene expression and protein synthesis. Other important activities attributable to AMPK are regulation of insulin synthesis and secretion in pancreatic islet b-cells and modulation of hypothalamic functions involved in the regulation of satiety. How these latter two functions impact obesity and diabetes will be discussed below.

STRUCTURE OF AMPK

The mammalian AMPK is a trimeric enzyme composed of a catalytic subunit and the non-catalytic and subunits. There are two genes encoding isoforms of both the and subunits (1, 2, 1 and 2) and three genes encoding isoforms of the subunit (1- 3). The 2 isoform is the subunit of AMPK found predominantly within skeletal and cardiac muscle, whereas, approximately equal distribution of both the 1 and 2 isoforms are present in hepatic AMPK. Within pancreatic islet -cells the 1 isoform predominates.

The N-terminal half of the a subunits contains a typical serine/threonine kinase catalytic domain. Interaction with the and subunits occurs via the C-terminal half of the subunits. The yeast AMPK subunits are lipid modified with myristic acid. Myristoylation may account for the membrane association mammalian AMPK. The core of the subunits have a glycogen-binding domain (GBD). This domain is closely related to the isoamylase N domain subfamily and weakly related to domains in the glycogen-targeting phosphatase subunits and several starch-binding proteins. The close proximity of AMPK to cellular glycogen stores allows it to rapidly affect changes in glycogen metabolism in response to changes in metabolic demands.

The subunits of AMPK have been shown to contain nucleotide binding sites with similarity to cystathionine -synthase (CBS) domains. Indeed, direct AMP-binding studies have shown that AMP is bound to the subunits by a pair of CBS domains. Of clinical significance is the observation that mutations in the CBS domains of the 2 subunit are associated with Wolff-Parkinson-White syndrome and familial hypertrophic cardiomyopathy.

REGULATION OF AMPK

AMPK is activated by phosphorylation by one or more

upstream AMPK kinases (AMPKKs). In the absence of phosphorylation there is no detectable activity of AMPK towards any substrates. Phosphorylation of AMPK occurs in the a subunit at threonine 172 (Thr-172) which lies in the activation loop.

One kinase activator of AMPK is Ca2+-calmodulin-dependent kinase kinase (CAMKK) which phosphorylates and activates AMPK in response to increased calcium. Recent evidence has demonstrated that the threonine kinase, LKB1, encoded by the Peutz-Jegher syndrome tumor suppressor gene, is required for activation of AMPK in response to stress. Loss of LKB1 activity in adult mouse liver leads to near complete loss of AMPK activity and is associated with hyperglycemia.

The hyperglycemia is, in part, due to an increase in the transcription of gluconeogenic genes. Of particular significance is the increased expression of the peroxisome proliferator-activated receptor- (PPAR-) coactivator 1 (PGC-1) which drives gluconeogenesis. Reduction in PGC-1 activity results in normalized blood glucose levels in LKB1-deficient mice. As the name implies, AMPK is also regulated by AMP. The effects of AMP are two-fold: a direct allosteric activation and making AMPK a poorer substrate for dephosphorylation. Because AMP affects both the rate of AMPK phoshorylation in the positive direction and dephosphorylation in the negative direction, the cascade is ultrasensitive. This means that a very small rise in AMP levels can induce a dramatic increase in the activity of AMPK. The activity of adenylate kinase, catalyzing the reaction shown below, ensures that AMPK is highly sensitive to small changes in the intracellular [ATP]/[ADP] ratio.

$$2\ ADP \rightarrow ATP + AMP$$

Negative allosteric regulation of AMPK also occurs and this effect is exerted by phosphocreatine. As indicated above, the subunits of AMPK have a glycogen-binding domain, GBD. In muscle, a high glycogen content represses AMPK activity and this is likely the result of interaction between the GBD and glycogen, although this has not been shown directly. As suggested above, the GBD of AMPK allows association of the enzyme with the regulation of glycogen metabolism by placing AMPK in close proximity to one of its substrates glycogen synthase. AMPK has also been shown to be activated by receptors that are coupled to phospholipase C- (PLC-) and by cytokines secreted by adipose tissue (adipocytokines) such as leptin and adiponectin.

TARGETS OF AMPK

The signaling cascades initiated by the activation of AMPK exert effects on glucose and lipid metabolism, gene expression and protein synthesis. These effects are most important for regulating metabolic events in the liver, skeletal muscle, heart, adipose tissue, and pancreas.

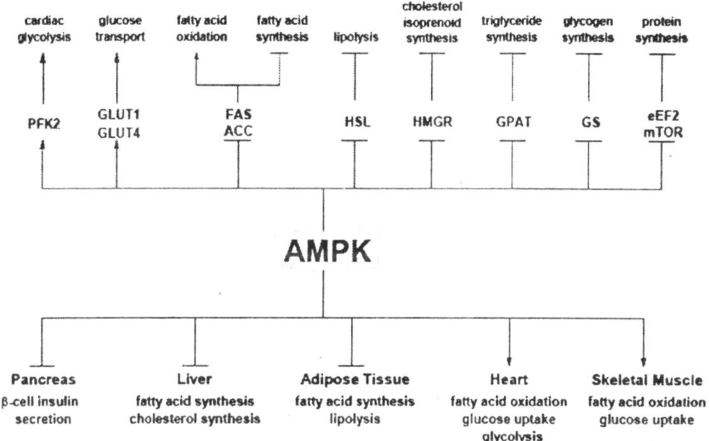

Demonstration of the central role of AMPK in the regulation of metabolism in response to events such as nutrient- or exercise-induced stress. Several of the known physiologic targets for AMPK are included as well as several pathways whose flux is affected by AMPK activation. Arrows indicate positive effects of AMPK, whereas, T-lines indicate inhibitory effects.

The uptake, by skeletal muscle, accounts for >70% of the glucose removal from the serum in humans. Therefore, it should be obvious that this event is extremely important for overall glucose homeostasis, keeping in mind, of course, that glucose uptake by cardiac muscle and adipocytes cannot be excluded from consideration. An important fact related to skeletal muscle glucose uptake is that this process is markedly impaired in individuals with type 2 diabetes. The uptake of glucose increases dramatically in response to stress (such as ischemia) and exercise and is stimulated by insulin-induced recruitment of glucose transporters to the plasma membrane, primarily GLUT4. Insulin-independent recruitment of glucose transporters also occurs in skeletal muscle in response to contraction (exercise).

The activation of AMPK plays an important, albeit not an exclusive, role in the induction of GLUT4 recruitment to the plasma membrane. In addition, there is some demonstration that AMPK may regulate glucose transport through GLUT1. Increased glucose uptake will result in an increase in glycolysis and ATP production. Under ischemic conditions in the heart the activation of AMPK leads to the phosphorylation and activation of phosphofructokinase-2, PFK-2 (6-phosphofructo-2-kinase). PFK-2 is one of the most potent regulators of the rate of flux through glycolysis and gluconeogenesis.

It is important to note that like many enzymes, there are multiple isoforms of PFK-2 and neither the liver nor the skeletal muscle isoforms contain the AMPK phosphorylation sites of the inducible form of PFK-2. Of particular significance is the fact that the inducible form of PFK-2 is commonly found in tumor cells and this may allow AMPK to play an important role in protecting tumor cells from hypoxic stress. Indeed, techniques for depleting AMPK in tumor cells have shown that these cells become sensitized to nutritional stress upon loss of AMPK activity.

Whereas, stress and exercise are powerful inducers of AMPK activity in skeletal muscle, additional regulators of its activity have been identified. Insulin-sensitizing drugs of the thiazolidinedione family (activators of PPAR-g, as well as the hypoglycemia drug metformin exert a portion of their effects through regulation of the activity of AMPK.

As indicated above, the activity of the AMPK activating kinase LKB1, is critical for regulation of gluconeogenic flux and consequent glucose homeostasis. The action of metformin in reducing blood glucose levels requires the activity of LKB1 in the liver for this function. Also, several adipocytokines either stimulate or inhibit AMPK activation: leptin and adiponectin have been shown to stimulate AMPK activation, whereas, resistin inhibits AMPK activation.

Within skeletal muscle and heart activation of AMPK leads to the phosphorylation and inhibition of acetyl-CoA carboxylase (ACC). This inhibition results in a drop in the level of malonyl-CoA which itself is an inhibitor of carnitine palmitoyltransferase I (CPT I).

With a drop in the inhibition of CPT I a concomitant increase in -oxidation of fatty acid will occur within the mitochondria. An increase in fatty acid oxidation, like increases in glycolysis, will lead to increases in ATP production.

In addiiton to ACC, AMPK has been shown to phosphorylate and thus regulate the activities of HMG-CoA reductase, (HMGR); hormone-sensitive lipase, (HSL); glycerophosphate acyltransferase, (GPAT); malonyl-CoA decarboxylase, (MCD); glycogen synthase, (GS) and creatine kinase, (CK). Therefore, the effects of AMPK activation are exerted on not only glucose homeostasis and fatty acid metabolism but overall energy homeostasis including glycogen metabolism, cholesterol metabolism and phosphocreatine metabolism.

Not only does activation of AMPK exerts effects on enzyme activity through phosphorylation, there are marked effects on the expression of a number of glycolytic and lipogenic enzymes in the liver and adipose tissue. Included in this list are the genes for the liver isoform of pyruvate kinase (L-PK), fatty acid synthase (FAS), and ACC. Activation of AMPK leads to a reduction in the level of SREBP a transcription factor that is a key regulator of the expression of numerous lipogenic enzymes.

Another transcription factor reduced in response to

AMPK activation is hepatocyte nuclear factor 4, HNF4 which is a member of the steroid/thyroid hormone superfamily. HNF4 in known to regulate the expression of several liver and pancreatic b-cell genes such as GLUT2, L-PK and preproinsulin.

Of clinical significance is that mutations in HNF4a are responsible for maturity-onset diabetes of the young, MODY-1. Recent evidence indicates that the newly discovered carbohydrate-response-element-binding protein (ChREBP) is a target for AMPK in the liver.

ChREBP is implicated in the transcriptional regulation of L-PK. The target of the thiazolidinedione class of drugs used to treat type 2 diabetes is the peroxisome proliferator-activated receptor, PPAR which itself may be a target for the action of AMPK. The transcription co-activator, p300, is phosphorylated by AMPK which inhibits interaction of p300 with not only PPAR but also the retinoic acid receptor, retinoid X receptor, and thyroid receptor. Another transcription factor target of AMPK is the forkhead protein, FKHR (now refered to as FoxO1a). FoxO1a is involved in the activation of glucose-6-phosphatase expression and, therefore, loss of FoxO1a activity in response to AMPK activation will lead to reduced hepatic output of glucose.

AMPK activation in response to hypoxia exerts effects on rates of protein synthesis. Hepatic translation elongation factor, eEF2 is a target for phosphorylation in repsonse to AMPK activation. AMPK phosphorylates and activates the kinase that phosphorylates eEF2 (eEF2K) leading to inhibition of protein synthesis. Another indirect substrate for AMPK that plays a role in protein synthesis is the mammalian target of rapamycin, mTOR. A detailed description of the role of mTOR in regulating protein synthesis.

RELEVANCE OF AMPK TO TYPE 2 DIABETES

An impairment in fuel metabolism occurs in obesity and this impairment is a leading pathogenic factor in type 2 diabetes. The insulin resistance associated with type 2 diabetes is most profound at the level of skeletal muslce as this is the primary site of glucose and fatty acid utilization. Therefore, an understanding of how to activate AMPK in skeletal muscle would offer significant pharmacologic benefits in the treatment of type 2 diabetes.

As indicated above, it has already been shown that metformin and the thiazolidinedione drugs exert some of their effects via activation of AMPK. In the non-pharmacologic context, activation of AMPK occurs in response to exercise, an activity known to have significant benefit for type 2 diabetics. Future targets for type 2 diabetes treatments will likely be those that can effect beneficial changes in the activity of AMPK.

LIPOPROTEINS

NTESTINAL UPTAKE OF LIPIDS

In order for the body to make use of dietary lipids, they must first be absorbed from the small intestine. Since these molecules are oils, they are essentially insoluble in the aqueous environment of the intestine. The solubilization (or emulsification) of dietary lipids is therefore accomplished by means of bile salts, which are synthesized from cholesterol in the liver and then stored in the gallbladder; they are secreted following the ingestion of fat.

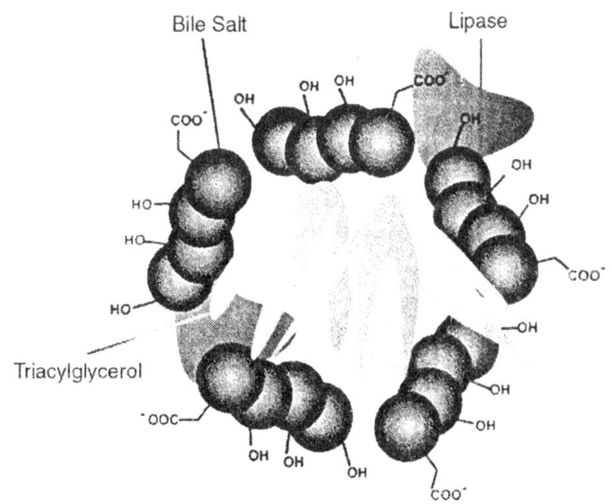

Mixed micelle formed by bile salts, triacylglycerols and pancreatic lipase.

The emulsification of dietary fats renders them accessible to pancreatic lipases (primarily lipase and phospholipase A2). These enzymes, secreted into the intestine from the pancreas, generate free fatty acids and a mixtures of mono- and diacylglycerols from dietary triacylglycerols. Pancreatic lipase degrades triacylglycerols at the 1 and 3 positions sequentially to generate 1,2-diacylglycerols and 2-acylglycerols. Phospholipids are degraded at the 2 position by pancreatic phospholipase A2 releasing a free fatty acid and the lysophospholipid. The products of pancreatic lipases then diffuse into the intestinal epithelial cells, where the re-synthesis of triacyglycerols occurs.

Dietary triacylglycerols and cholesterol, as well as triacylglycerols and cholesterol synthesized by the liver, are solubilized in lipid-protein complexes. These complexes contain triacylglycerol lipid droplets and cholesteryl esters surrounded by the polar phospholipids and proteins identified as apolipoproteins. These lipid-protein complexes vary in their content of lipid and protein.

INTERMEDIATE DENSITY LIPOPROTEINS, IDLS

IDLs are formed as triacylglycerols are removed from VLDLs. The fate of IDLs is either conversion to LDLs or direct uptake by the liver.

Conversion of IDLs to LDLs occurs as more triacylglycerols are removed. The liver takes up IDLs after they have interacted with the LDL receptor to form a complex, which is endocytosed by the cell.

For LDL receptors in the liver to recognize IDLs requires the presence of both apo-B-100 and apo-E (the LDL receptor is also called the apo-B-100/apo-E receptor). The importance of apo-E in cholesterol uptake by LDL receptors has been demonstrated in transgenic mice lacking functional apo-E genes. These mice develop severe atherosclerotic lesions at 10 weeks of age.

LOW DENSITY LIPOPROTEINS, LDLS

The cellular requirement for cholesterol as a membrane component is satisfied in one of two way: either it is synthesized de novo within the cell, or it is supplied from extra-cellular sources, namely, chylomicrons and LDLs. As indicated above, the dietary cholesterol that goes into chylomicrons is supplied to the liver by the interaction of chylomicron remnants with the remnant receptor.

In addition, cholesterol synthesized by the liver can be transported to extra-hepatic tissues if packaged in VLDLs. In the circulation VLDLs are converted to LDLs through the action of lipoprotein lipase. LDLs are the primary plasma carriers of cholesterol for delivery to all tissues.

The exclusive apolipoprotein of LDLs is apo-B-100. LDLs are taken up by cells via LDL receptor-mediated endocytosis, as described above for IDL uptake. The uptake of LDLs occurs predominantly in liver (75/%), adrenals and adipose tissue. As with IDLs, the interaction of LDLs with LDL receptors requires the presence of apo-B-100. The endocytosed membrane vesicles (endosomes) fuse with lysosomes, in which the apoproteins are degraded and the cholesterol esters are hydrolyzed to yield free cholesterol. The cholesterol is then incorporated into the plasma membranes as necessary. Excess intracellular cholesterol is re-esterified by acyl-CoA-cholesterol acyltransferase (ACAT), for intracellular storage. The activity of ACAT is enhanced by the presence of intracellular cholesterol.

Insulin and tri-iodothyronine (T3) increase the binding of LDLs to liver cells, whereas glucocorticoids (e.g., dexamethasone) have the opposite effect. The precise mechanism for these effects is unclear but may be mediated through the regulation of apo-B degradation. The effects of insulin and T3 on hepatic LDL binding may explain the hypercholesterolemia and increased risk of athersclerosis that have been shown to be associated with uncontrolled diabetes or hypothyroidism.

An abnormal form of LDL, identified as lipoprotein-X (Lp-X), predominates in the circulation of patients suffering from lecithin-cholesterol acyl transferase deficiency or cholestatic liver disease. In both cases there is an elevation in the level of circulating free cholesterol and phospholipids.

HIGH DENSITY LIPOPROTEINS, HDLS

HDLs are synthesized de novo in the liver and small intestine, as primarily protein-rich disc-shaped particles. These newly formed HDLs are nearly devoid of any cholesterol and cholesteryl esters. The primary apoproteins of HDLs are apo-A-I, apo-C-I, apo-C-II and apo-E. In fact, a major function of HDLs is to act as circulating stores of apo-C-I, apo-C-II and apo-E.

HDLs are converted into spherical lipoprotein particles through the accumulation of cholesteryl esters. This accumulation converts nascent HDLs to HDL2 and HDL3. Any free cholesterol present in chylomicron remnants and VLDL remnants (IDLs) can be esterified through the action of the HDL-associated enzyme, lecithin-cholesterol acyl transferase, LCAT. LCAT is synthesized in the liver and so named because it transfers a fatty acid from the C-2 position of lecithin to the C-3-OH of cholesterol, generating a cholesteryl ester and lysolecithin. The activity of LCAT requires interaction with apo-A-I, which is found on the surface of HDLs.

Cholesterol-rich HDLs return to the liver, where they are endocytosed. Hepatic uptake of HDLs, or reverse cholesterol transport, may be mediated through an HDL-specific apo-A-I receptor or through lipid-lipid interactions. Macrophages also take up HDLs through apo-A-I receptor interaction. HDLs can then acquire cholesterol and apo-E from the macrophages; cholesterol-enriched HDLs are then secreted from the macrophages. The added apo-E in these HDLs leads to an increase in their uptake and catabolism by the liver. HDLs also acquire cholesterol by extracting it from cell surface membranes. This process has the effect of lowering the level of intracellular cholesterol, since the cholesterol stored within cells as cholesteryl esters will be mobilized to replace the cholesterol removed from the plasma membrane.

The cholesterol esters of HDLs can also be transferred to VLDLs and LDLs through the action of the HDL-associated enzyme, cholesterol ester transfer protein (CETP, also identified as apo-D). This has the added effect of allowing the excess cellular cholesterol to be returned to the liver through the LDL-receptor pathway as well as the HDL-receptor pathway.

LDL RECEPTORS

LDLs are the principal plasma carriers of cholesterol delivering cholesterol from the liver (via hepatic synthesis of VLDLs) to peripheral tissues, primarily the adrenals and adipose tissue. LDLs also return cholesterol to the liver. The cellular uptake of cholesterol from LDLs occurs following the interaction of LDLs with the LDL receptor (also called the apo-B-100/apo-E receptor). The sole apoprotein present in LDLs is apo-B-100, which is required for interaction with the LDL receptor.

The LDL receptor is a polypeptide of 839 amino acids that spans the plasma membrane. An extracellular domain is responsible for apo-B-100/apo-E binding. The intracellular domain is responsible for the clustering of LDL receptors into regions of the plasma membrane termed coated pits. Once LDL binds the receptor, the complexes are rapidly internalized (endocytosed). ATP-dependent proton pumps lower the pH in the endosomes, which results in dissociation of the LDL from the receptor. The portion of the endosomal membranes harboring the receptor are then recycled to the plasma membrane and the LDL-containing endosomes fuse with lysosomes. Acid hydrolases of the lysosomes degrade the apoproteins and release free fatty acids and cholesterol. As indicated above, the free cholesterol is either incorporated into plasma membranes or esterified (by ACAT) and stored within the cell.

The level of intracellular cholesterol is regulated through cholesterol-induced suppression of LDL receptor synthesis and cholesterol-induced inhibition of cholesterol synthesis. The increased level of intracellular cholesterol that results from LDL uptake has the additional effect of activating ACAT, thereby allowing the storage of excess cholesterol within cells. However, the effect of cholesterol-induced suppression of LDL receptor synthesis is a decrease in the rate at which LDLs and IDLs are removed from the serum. This can lead to excess circulating levels of cholesterol and cholesteryl esters when the dietary intake of fat and cholesterol exceeds the needs of the body. The excess cholesterol tends to be deposited in the skin, tendons and (more gravely) within the arteries, leading to atherosclerosis.

SIGNIFICANCES OF LIPOPROTEIN METABOLISM

Fortunately, few individuals carry the inherited defects in lipoprotein metabolism that lead to hyper- or hypolipoproteinemias. Persons suffering from diabetes mellitus, hypothyroidism and kidney disease often exhibit abnormal lipoprotein metabolism as a result of secondary effects of their disorders. For example, because lipoprotein lipase (LPL) synthesis is regulated by insulin, LPL deficiencies leading to Type I hyperlipoproteinemia may occur as a secondary outcome of diabetes mellitus. Additionally, insulin and thyroid hormones positively affect hepatic LDL-receptor interactions; therefore, the hypercholesterolemia and increased risk of athersclerosis associated with uncontrolled diabetes or hypothyroidism is likely due to decreased hepatic LDL uptake and metabolism.

Of the many disorders of lipoprotein metabolism, familial hypercholesterolemia (FH) may be the most prevalent in the general population. Heterozygosity at the FH locus occurs in 1:500 individuals, whereas, homozygosity is observed in 1:1,000,000 individuals. FH is an inherited disorder comprising four different classes of mutation in the LDL receptor gene.

The class 1 defect (the most common) results in a complete loss of receptor synthesis. The class 2 defect results in the synthesis of a receptor protein that is not properly processed in the Golgi apparatus and therefore is not transported to the plasma membrane. The class 3 defect results in an LDL receptor that is incapable of binding LDLs.

The class 4 defect results in receptors that bind LDLs but do not cluster in coated pits and are, therefore, not internalized.

FH sufferers may be either heterozygous or homologous for a particular mutation in the receptor gene. Homozygotes exhibit grossly elevated serum cholesterol (primarily in LDLs). The elevated levels of LDLs result in their phagocytosis by macrophages. These lipid-laden phagocytic cells tend to deposit within the skin and tendons, leading to xanthomas. A greater complication results from cholesterol deposition within the arteries, leading to atherosclerosis, the major contributing factor of nearly all cardiovascular diseases.

PHARMACOLOGIC INTERVENTION

Drug treatment to lower plasma lipoproteins and/or cholesterol is primarily aimed at reducing the risk of athersclerosis and subsequent coronary artery disease that exists in patients with elevated circulating lipids. Drug therapy usually is considered as an option only if non-pharmacologic interventions (altered diet and exercise) have failed to lower plasma lipids.

- Mevinolin, Mevastatin, Lovastatin: These drugs are fungal HMG-CoA reductase inhibitors. The net result of treatment is an increased cellular uptake of LDLs, since the intracellular synthesis of cholesterol is inhibited and cells are therefore dependent on extracellular sources of cholesterol. However, since mevalonate (the product of the HMG-CoA reductase reaction) is required for the synthesis of other important isoprenoid compounds besides cholesterol, long-term treatments carry some risk of toxicity.
- Nicotinic acid: Nicotinic acid reduces the plasma levels of both VLDLs and LDLs by inhibiting hepatic VLDL secretion, as well as suppressing the flux of FFA release from adipose tissue by inhibiting lipolysis. Because of its ability to cause large reductions in circulating levels of cholesterol, nicotinic acid is used to treat Type II, III, IV and V hyperlipoproteinemias.

- Clofibrate, Gemfibrozil, Fenofibrate: These compounds are derivatives of fibric acid and promote rapid VLDL turnover by activating lipoprotein lipase. They also induce the diversion of hepatic free fatty acids from esterification reactions to those of oxidation, thereby decreasing the liver's secretion of triacylglycerol- and cholesterol-rich VLDLs.
- Probucol: Probucol increases the rate of LDL metabolism and may block the intestinal transport of cholesterol. The net result is a significant reduction in plasma cholesterol levels.
- Cholestyramine or colestipol (resins): These compounds are nonabsorbable resins that bind bile acids which are then not reabsorbed by the liver but excreted. The drop in hepatic reabsorption of bile acids releases a feedback inhibitory mechanism that had been inhibiting bile acid synthesis. As a result, a greater amount of cholesterol is converted to bile acids to maintain a steady level in circulation. Additionally, the synthesis of LDL receptors increases to allow increased cholesterol uptake for bile acid synthesis, and the overall effect is a reduction in plasma cholesterol. (This treatment is ineffective in homozygous FH patients, since they are completely deficient in LDL receptor

MEMBRANE LIPIDS AND BILAYER MEMBRANES

The simpler lipids are used primarily for energy storage by living organisms, but more complex lipids are used to form biological membranes. The most important and abundant class of biological membrane lipids is the class of phospholipids. Most phospholipids are derived from glycerol and so they are called phosphoglycerides. The amino alcohol sphingosine gives rise to sphingomyelin, the only significant membrane phospholipid which is not a phosphoglyceride.

In phosphoglycerides, the hydroxy groups at C-1 and C-2 of glycerol are esterified to fatty acids while the hydroxy group at C-3 of glycerol is esterified to phosphoric acid; this forms diacylglycerol-3-phosphates. The two fatty acids found in a diacylglycerol-3-phosphate molecule need not be the same. Diacylglycerol-3-phosphates are only a minor constituent of membranes. The major constituents of membranes are derivatives of diacylglycerol-3-phosphates in which the esterified phosphate also forms an ester linkage with an alcohol. The alcohols commonly found in phosphoglycerides include glycerol, ethanolamine, and choline.

The glycolipids are lipids which contain sugars. Glycolipids are derived from sphingosine, as is the phospholipid sphingomyelin. Glycolipids contain no esterified phosphate.

Both phospholipids and glycolipids are amphipathic molecules, that is, molecules in which one end is hydrophobic and one end is hydrophilic. The fatty acid chains form the hydrophobic end of the molecule while the polar glycerol-phosphate-alcohol or sphingosine-sugar portion forms the hydrophilic end of the molecule.

BILAYER MEMBRANES

When amphipathic molecules are placed in aqueous solution, their hydrophobic tails attempt to orient themselves toward each other. The two ways in which they do so are to form small spherical micelles or to form a planar lipid bilayer. A micelle is a small structure, generally less than two micrometre in diametre, while a lipid bilayer typically has a thickness of about 0.5 micrometre and can have an area of several square millimetres. Sonication or other treatment of lipids can produce spherical liposomes, or lipid vesicles, which contain aqueous solution within an enclosing lipid bilayer. Liposomes are considerably larger than micelles.

Both liposomes and planar bilayer membranes can be prepared from simple solutions of phospholipids such as the diacylglycerol-3-phosphatidyl cholines. Studies of such artificially prepared membranes have given us much insight into the functions and operation of membranes in living cells.

Simple bilayer membranes are highly permeable to water molecules, while ions such as sodium ion and potassium ion can traverse them only more slowly (by nine orders of magnitude!). For small molecules, the rates of permeation through simple bilayer membranes increase with their solubility in nonpolar solvents relative to their solubility in water. These observations strongly suggest that transfer of substances through a membrane requires desolvation, transfer through the anhydrous membrane interior, and then resolvation on the other side of the membrane.

The differences in transport behaviour between biological membranes and simple bilayer membranes are now believed to be due to the incorporation of protein molecules or other specifically transporting molecules directly into or onto the surface of a bilayer cell membrane. Considerable research in biology and in chemistry is being directed toward the transport properties of membranes.

LIPIDS AND FATTY ACIDS

Lipids are the generic names assigned to a group of fat soluble compounds found in the tissues of plants and animals,: and are broadly classified as:

- Fats,
- Phospholipids,
- Sphingomyelins,
- Waxes,
- Sterols.

Fats are the fatty acid esters of glycerol and are the primary energy depots of animals. These are used for long-term energy requirements during periods of extensive exercise or during periods of inadequate food and energy intake. Fish have the unique capability of metabolizing these compounds readily and, as a result, can exist for long periods of time under conditions of food deprivation. A typical example is the many weeks of migration by salmon in their return upstream to spawn; stored lipid deposits are burned for fuel to enable body processes to continue during the strenuous journey.

Phospholipids are the esters of fatty acids and phosphatidic acid. These are the main constituent lipids of cellular membranes allowing the membrane surfaces to be hydrophobic or hydrophylic depending on the orientation of the lipid compounds into the intra or extracellular spaces. Sphingomyelins are the fatty acid esters of sphingosine and are present in brain and nerve tissue compounds. Waxes are fatty acid esters of long-chain alcohols. These compounds can be metabolized for energy and to impart physical and chemical characteristics through the stored lipids of some plant and several animal compounds. Sterols are polycyclic, long-chain alcohols and function as components of several hormone systems, especially in sexual maturation and sex-related physiological functions.

Fatty acids can exist as straight chain or branch chain components; many of the fish fats contain numerous unsaturated double bonds in the fatty acid structures. A short bond designation for. fatty acids will be used throughout where the number identifies the position of the first double bond counting from the methyl end. Linolenic acid would be written 18:3ω3.

The first number identifies the number of carbons; the second number, the number of double bonds; and the last number, the position of the double bonds. Many reviews of fish nutrition have been published which contain information on lipid requirements. Most work on lipid requirements of fish has been with salmonids. Rainbow trout have an essential fatty acid (EFA) requirement for the linolenic of ω3[1] series rather than for linolenic or ω6 as required by most mammals. The main emphasis on lipid requirements has been on EFA and on the energy value of lipids.

FATTY ACID COMPOSITION OF FISH

ENVIRONMENTAL INFLUENCES

Salinity

The difference between fatty acid compositions of marine

and freshwater fish has been noted by several authors. Some examples of fatty acid patterns are given in Table 1. Although these fish lipids are higher in ω 3 fatty acids, it is clear that freshwater fish have higher levels of ω 6 fatty acids than marine species.

The average ω6/ω3 ratios are 0.37 and 0.16 for freshwater and marine fish, respectively. Fish in general contain more ω 3 than ω 6 polyunsaturated fatty acids (PUFA) and should have a higher dietary requirement for ω 3 PUFA; thus the dietary EFA requirement of marine fish for ω 3 PUFA may be higher than that of freshwater fish. The same type of difference in the ω6/ω3 ratio between freshwater and seawater is seen when some species of fish migrate from oceans to streams or vice versa.

The PUFA ratio of sweet smelt (Plecoglosus altivelis) changes drastically in only one month as they migrate from the sea to a freshwater river. A similar but reverse change occurs in the masu salmon (Oncorhynchus masu) as they migrate from freshwater to seawater. Even within the same species of fish, the salinity of the water seems to cause a dramatic change in the fatty acid pattern. The difference between marine and freshwater fish may be due simply to differences in the fatty acid content in the diet or it may be related to a specific requirement in fish related to physiological adaptations to the environments.

The phospholipids are generally considered to be structural or functional lipid, being incorporated to a large extent in the membrane structure of cell and subcellular particles. The triglycerides are more often storage lipids and reflect the fatty acid composition of the diet to a greater extent than do the phospholipids.

In Table the fatty acid compositions of the triglyceride and phospholipid fractions of fish lipids are presented. It can be seen that the effect of changing environment on the fatty acid composition of the phospholipid is as great in the case of salmon, and considerably greater in the case of the sweet smelt, than it is on the triglyceride composition. Rainbow trout on diets containing either corn oil, which is high in ω6 but low in ω6 PUFA, showed a higher mortality and growth reduction in seawater than in freshwater over the twelve-week feeding period.

Temperature

There are several other factors besides the salinity of the water which affect the fatty acid composition and especially the PUFA of fish. In Tables 1 and 2 it can be seen that the salmonids, even in freshwater, tend to have a higher total PUFA of the 20 and 22 carbon chain length, and a lower ω6/ω3 ratio than the other fish. The salmonids are mostly cold-water fish.

The fatty acids from a number of marine animals from temperate and arctic waters show some significant differences in the general pattern; unfortunately analysis included fatty acids longer than 20:1. There are a number of other experiments demonstrating the effect of environmental temperature on fatty acid composition of aquatic animals. The general trend toward higher content of long chain PUFA at lower temperatures is quite clear.

The ω6/ω3 ratio decreases with a decrease in temperature (Table 3). If the trends in fatty acid composition can be taken as clues to the EFA requirements of fish, the ω3 requirement would be greater for fish raised at lower temperatures. Fish raised in warmer waters, such as common carp, channel catfish, and tilapia may do better with a mixture of ω6 and ω3 fatty acids.

Effects of Diet

Some of the fatty acid compositions listed in Table 3 may be seriously affected by the dietary lipids. The mosquito fish and guppies were fed trout pellets which had an ω6/ω3 ratio of 2.75. The catfish were fed diets supplemented with either beef tallow or menhaden oil, with ω6/ω3 ratios of 18.13 and 0.15, respectively.

These fish were able to alter the dietary ω6/ω3 ratio in favour of ω3 fatty acid incorporation into the flesh lipids even at the highest temperature. Commercially available trout pellets are often low in ω3 PUFA and high in ω6 fatty acids. It is important not to ignore the effect of dietary lipid composition on fatty acid composition of fish fed artificial diets. It is clear from the data in Table 3 that the ω6/ω3 ratio of the fish lipids is greatly affected by the ω6/ω3 ratio of the dietary lipids. When the dietary ratio is very high in ω6 fatty acids supplied by animal lard or vegetable oils, there is a tendency for fish to alter the ratio of PUFA incorporated in favour of ω3 fatty acids. When the dietary oil is a fish oil high in to 3 fatty acids, there is little change in the ω6/ω3 ratio of lipids incorporated into the fish. This is further suggestive evidence of an EFA requirement of fish for ω3 PUFA.

Seasonal Variation

Seasonal variations in the fatty acid composition of fish species have often been reported. Seasonal changes have been observed in total lipid and iodine values of herring oils. The iodine value or degree of unsaturation of the oil was minimal in April and maximal in June. The great increase in unsaturation corresponded to the onset of feeding in spring. The absence of a gas liquid chromatograph (GLC) at the time precluded identification of changes in individual fatty acids. Flesh and viscera lipid content of the sardine Sardinops melanosticta vary from 3.9 to 10.77 per cent and from 10.9 to 38.3 per cent, respectively.

The fatty acids of principal interest with respect to EFA metabolism are 20:4ω6, 20:5ω3, and 22:6ω3. There was considerable variation in all of these fatty acids in both neutral and polar lipid from both tissues.

In the flesh, the 20:4ω6 was consistently higher in the neutral lipid than in the polar lipid. The total 20:5 3 plus 22:6 3 was consistently higher in polar lipid than in the neutral lipid. Thus, in spite of the major fluctuations in fatty acids caused by changes in diet and temperature throughout the seasons, there was a consistent preferential incorporation of PUFA of the ω3 series into the polar or phospholipid fraction of the lipids.

One of the best clues to the EFA requirements of a species can be gained from the fatty acid composition of the lipids incorporated into the offspring or egg. The act of reproduction or spawning also has a significant effect on the seasonal fluctuation of lipids in fish.

Fatty acid composition of fish egg lipids is probably distinctive for each species and contains increased levels of 16:0, 20:4ω6, 20:5ω3 and 22:6ω3 compared to the liver lipids of the same female fish (Ackman, 1967). Elevated levels of 16:0, 20:5 3, and 22:6ω3 and reduced 18:1 in the ovary occurred compared to mesenteric fat of Pacific sardine fed a natural copepod diet.

The blood fatty acids of the sardine fed the natural diet were similar to those of the ovary. When the sardines were fed trout food, both the blood and mesenteric fat responded to the diet with elevated 18:2ω6 and reduced 20:5ω3 arid 22:6ω3.

The effect of the diet on ovary fatty acid content was considerably less, as relatively high levels of 20:5ω3 and 22:6ω3 were retained. The ovary lipids of the sweet smelt show an increase in 16:0, and a reduction in the PUFA, especially in the phospholipids, compared to the lipids from the flesh of fish caught at the same time of year. The ω6/ω3 ratio of the ovary was lower than that of the flesh lipids, 0.21 and 0.17 for the ovary compared to 0.31 and 0.20 for the triglycerids and phospholipids of the flesh, respectively.

The hatchability of eggs from common carp fed several different formulated feeds is greatly reduced when the 22:6ω3 of the egg lipids is less than 10 per cent. Further, the muscle, plasma, and erythrocyte fatty acid compositions are more affected by dietary lipid than those of the eggs. The EFA requirements of a number of species of fish have been investigated in nutritional studies.

The fish themselves have given ample evidence for EFA preference by the types of fatty acids they incorporate into their lipids. Fish, in general, tend to utilize ω3 over ω6. This is especially observed when the dietary lipids are high in ω6, as the fish tend to alter the ω6/3 ratio toward the ω3 fatty

acids in the tissue lipids. The lipids of the egg must satisfy the EFA requirement of the embryo until it is able to feed. The fatty acid composition data suggest that the ω3 requirement is greater in seawater than in freshwater and higher in cold water than in warm water.

LIPID COMPOSITION AND DIETARY LIPID

Detailed information is still lacking on the dietary lipid requirements of many species of fish, but there is an abundance of information on the fatty acid composition of fish oils. Information on the lipid composition of fish can be used to make some guesses about dietary lipid requirements.

Linolenic acid (18:3ω3) resulted in some sparing action and growth promotion in rats, and fatty acids of the ω6 EFA prevented all of the EFA-deficiency symptoms.

Research with homeothermic land-dwelling animals showed that the ω6 series of fatty acids are the"essential fatty acids", while the ω3 series are considered to be non-essential or only have a partial sparing action on EFA-deficiency. The ω6 series of fatty acids have been shown to be essential to enough species that it began to become accepted that these are the essential fatty acids for all animals.

It was assumed by many that fish also required ω6 fatty acids. Many researchers began by supplementing fish diets with vegetable oils, such as corn, peanut, or sunflower oil, which were rich in linoleic acid. The main sympton observed during the development of EFA deficiency in chinook salmon fed fat-free diets was a marked depigmentation that can be prevented by addition of 1 per cent trilinolein, but not by 0.1 per cent linolenic acid.

Although the ω6 fatty acids are considered to be essential, one of the general characteristics of fish oils is the low levels of ω6 series fatty acids and the higher levels of ω3 type fatty acids. There is evidence that polyunsaturated fatty acids (PUPA) of the ω3 series, which are present in relatively large concentrations in fish oil, play the role of essential fatty acid for fish.

POLYUNSATURATED FATTY ACID

When a test diet containing 13 per cent corn oil and 2 per cent cod-liver oil was fed to rainbow trout, subsequent deletion of the cod-liver oil from the diet depressed growth and produced some kidney degeneration which might be attributed to a lack of sufficient ω3 PUFA present in significant quantities in cod-liver oil (McLaren et al., 1947).

Dietary fish oil is superior to corn oil in promoting growth of rainbow trout (Salmo gairdneri) and the yellow-tail (Seriola guingueradiata). Dietary linolenic acid or ethyl linolenate (18:3ω3) gives a positive growth response for rainbow trout which may be attributed to a dietary requirement for ω3 fatty acids.

ESSENTIAL FATTY ACID REQUIREMENTS OF FISH

One of the most widely accepted theories explaining the presence of such high levels of 20:5ω3 and 22:6ω3 fatty acids in fish oils is related to the effect of unsaturation on the melting point of a lipid.

The greater degree of unsaturation of fatty acids in the fish phospholipids allows for flexibility of cell membrane at lower temperatures. The ω3 structure allows a greater degree of unsaturation than the ω6 or ω9. This theory is consistent with the fact that cold water fish have a greater nutritional requirement for ω3 fatty acids, while the EFA requirement of some warm water fish can be satisfied by a mixture of ω6 plus ω3.

Rainbow Trout

Rainbow trout, a cold water fish, requires ω3 fatty acids as EFA in the diet. The EFA requirement can be met by 1 per cent 18:3ω3 in the diet. Inclusion of 18:2ω6 in the diet may result in some improvement in growth and feed conversion compared to EFA deficient diets; however, the ω6 fatty acids will not prevent some EFA deficiency symptoms such as the"shock syndrome". Although it is clear that rainbow trout require ω3 fatty acids, it remains to be shown conclusively whether some dietary level of ω6 fatty acid is essential.

In all the above studies with rainbow trout, dietary 18:2ω6 or 18:3ω3 were readily converted to C-20 and C-22 PUFA of the same series, and 18:3ω3 or 22:6ω3 had similar EFA value for rainbow trout.

Either 20:5ω3 or 22:6ω3 is superior to 18:3ω3 in an EFA value for rainbow trout, and the former two fatty acids in combination are superior to either alone. This is consistent with data for mammals, where 20:4ω6 has higher EFA value than 18:2ω6. The superior nutritional value of C-20 and C-22 carbon ω3-PUFA is further supported by the excellent growth promoting effects of dietary fish oils such as pollock liver oil and salmon oil for rainbow trout.

Channel Catfish

One of the most important warm water fish in North America is the channel catfish (Ictalurus punctatus). The quantitative EFA requirement of the catfish has not yet been determined. However, the evidence is convincing that the ω3 requirement is not as great as that of rainbow trout. Analysis of fatty acids of lipids from catfish purchased at five processing plants showed very low levels of 20:4ω6, 20:5ω3, and 22:6ω3; 0.8 - 5.5, 0.2 - 1.3, and 0.6 - 6.1 per cent of the total fatty acids, respectively.

It was shown that corn oil added to a semipurified casein based diet initially resulted in a positive growth response and protein sparing, but later growth inhibition was observed. The apparent repressive effects of corn oil may be due to its 18:2ω6 content since 20:5ω3 and 22:6ω3 present in menhaden oil had no apparent detrimental effects. The growth suppressing effects of 18:2ω6 were also noted when 3 per cent corn oil was added to 3 per cent beef tallow and 3 per cent menhaden oil. The growth suppression caused by unsaturated fatty acids does not appear to be limited to ω6 fatty acids. Linseed oil (high in 18:3ω3) in the diet of catfish resulted in growth suppression similar to that caused by corn oil compared with dietary beef tallow, olive oil and menhaden oil.

The Common Carp

The picture for another warm water fish, the common carp (Cyprinus carpio) is much clearer than that for the channel catfish. This fish has an EFA requirement for both ω3 and ω6 fatty acids. The best weight gains and feed conversions are obtained in fish receiving a diet containing both 1 per cent 18:2ω6 and 1 per cent 18:3ω3. With the carp, 20:5ω3 and 22:6ω3 at 0.5 per cent of the diet are superior to 1 per cent of 18:3u3. Carp fed a fat-free, or EFA deficient, diet incorporated high levels of 20:3 9 in their lipids, especially in the phospholipids.

The Eel

The eel (Anguilla japonica), another warm water fish, has a requirement for both ω3 and ω6 fatty acids. Corn oil (high in ω6) and cod liver oil (high in ω3)in a 2:1 mixture are most favourable for the growth of eels. The eel requires ω6 and ω3 in the same proportion as the carp, but at a lower level in the diet; namely, 0.5 per cent of each, rather than 1.0 per cent of each PUFA.

The Plaice

The plaice becomes depleted of both ω3 and ω6 PUFA when fed a fat-free diet. The addition of 12:0 and 14:0 to the diet result in the synthesis of saturated and monoenoic fatty acids of chain lengths up to C18; however, increased levels of 20:3ω9 noted in trout and mammals have not been reported in plaice. Plaice fed dietary 18:2ω6 and 18:3ω3 will not produce significant amounts of 20:4 ω6, 20:5ω3, or 22:6ω3.

The Turbot

The growth of turbot (Scophthalmus matimus) is much better with ω3 PUFA than with ω6 or saturated fat (hydrogenated coconut oil) in the diet. The turbot also appears to be unable to convert dietary 18:2ω6 to 20:4ω6 when fed corn oil, or to convert endogenous 18:1ω9 to 20:3ω9 when fed the EFA deficient diet.

Although it appears to have an EFA requirement for

ω3 fatty acids such as are present in cod liver oil, this requirement is not satisfied by 18:3ω3. The chain elongation and desaturation of 18:lω9, 18:2ω6, or 18:3ω3 has been found to be very limited (3-15 per cent) in turbot compared to the rainbow trout where 70 per cent of the 18:3ω3 was converted to 22:6ω3. The required level of long-chain ω3 fatty acids for turbot is at least 0.8 per cent of the diet.

The Red Sea Bream

The red sea bream (Chrysophyrys major) grows better when the dietary lipid is of marine origin (pollock residual oil) rather than a vegetable oil (such as corn oil). The EFA requirement of the red sea bream is not satisfied by either linoleic acid of corn oil or supplemented linolenate. A mixture of 20:5ω3 and 22:6ω3 supplemented to the corn oil diet has been shown to be effective in improving growth and condition of these fish.

Thus, even in warm water, marine fish seem to require not just ω3 fatty acids but 0)3 fatty acids of 20 to 22 carbon-chain length. A direct correlation between feed efficiency and the 18:1 level in the lipids of the red sea bream has been postulated.

Other Species

Among warm water marine fish, mullet and fundulus possess the ability to chain, elongate, and desaturate 18:2ω6 or 18:3ω3 PUFAs. The process is, however, inhibited in fundulus by high levels (about 5 per cent) of these PUFAs of 18:2ω6 or 18:3ω3 in the diet.

FATTY ACID METABOLISM IN FISH

It appears that high levels of 18-carbon ω6 or ω3 fatty acids inhibit the synthesis and metabolism of 18:lω9. It is interesting to note that the channel catfish, which also exhibits negative growth response to dietary 18:2ω6 or 18:3ω3, incorporates very high levels of 18:1 into its body lipids. The inclusion of either 18:2ω6 or 18:3ω3 in the diet reduces the levels of 18:1 fatty acids in body lipids. A similar reduction has also been observed in red sea bream liver phospholipid when either of the PUFAs is added to the diet.

The competitive inhibition of chain elongation and desaturation of members from one series of fatty acids for members of another series is well established, with ω3 > 6 > 9 being the usual order of potency for inhibition.

The pathways of fatty acid metabolism have been reviewed. Fish are able to synthesize, de novo from acetate, the even-chain, saturated fatty acids, as shown. Radio tracer studies have shown that fish can convert 16:0 to the ω7 monoene and 18:0 to the ω9 monoene. The ω5, 11 and ω3 monoenes are proposed based on the identification of these isomers in the monoenes of herring oil.

Fish are unable to synthesize any fatty acids of the w6 and u3 series unless a precursor with this structure is present in the diet. Fish are able to desaturate and elongate fatty acids of the w9, 6, or w3 series as outlined in Fig. There is competitive inhibition of the elongation desaturation of fatty acids of one series by members of the other series. The w3 fatty acids are the most potent inhibitors, the w9 are the least.

The ability to elongate and desaturate fatty acids is not the same in all species of fish, as was noted earlier. The turbot was able to desaturate and elongate only 3-15 per cent of 18:1w9, 18:2w6, or 18:3w3, when given the C14 labelled fatty acid; in the rainbow trout, 70 per cent of the label from 18:3w3 (C14) was found in 22:6w3.

The essential fatty acids are not unique in their ability to supply energy. The -oxidation of fatty acids in fish is basically the same as in mammals. The EFA and saturated and monoenoic fatty acids are all equally utilized by fish for energy production.

Increased swelling rates of liver mitochondria occur in rainbow trout fed diets deficient in ω3 fatty acids. It is possible that EFA plays an important role in the permeability as well as the plasticity of membranes. The role of ω3 fatty acids in membrane permeability may be one of the factors accounting for differences in content of this family of fatty acids between freshwater and marine fish.

Fish mitochondria with high levels of the ω3 PUFA and very low levels of ω6 fatty acids are very similar to mammalian mitochondria with respect to cytochrome content, -oxidation of fatty acids, operation of the tricarboxylic acid cycle, electron transport, and oxidative phosphorylation. The ω3 PUFA may play the same role in fish that the ω6 fatty acids play in rats. The EFA play another role in the mitochondria. In addition to their importance in membrane structure, the EFA are important in some enzyme systems.

Unsaturated fatty acids play an important role in the transportation of other lipids. It has been repeatedly shown that feeding PUFA will lower the cholesterol levels in animals with above-normal blood lipid and cholesterol levels.

Fish oils are more effective in lowering cholesterol levels than are most dietary lipids.

The major portion of the fatty acids absorbed across the intestinal mucosa are transported as protein-lipid complexes stabilized by phospholipids. The low body temperature in fish probably results in a greater importance for unsaturation in transport of lipids than in homeothermic animals.

NEGATIVE ASPECTS OF LIPIDS IN FISH NUTRITION

The requirement by fish for PUFA of the ω3 series creates problems with respect to feed storage. These types of fatty acids are very labile on oxidation. The products of lipid oxidation may react with other nutrients such as proteins, vitamins, etc., and reducing the available dietary levels or the oxidation products may be toxic. The effect of oxidized lipids on dietary proteins, enzymes and amino acids have been demonstrated.

The use of oxidized menhaden oil in the diets of swine and rats caused decreased appetite, reduced growth, yellowish-brown pigmentation of depot fat, and decreased haemoglobin and haematocrit levels.

The negative effects of the oxidized fish oils were reversed by the addition of alpha-tocopherol acetate or ethoxyoquin to the diet.

Much of the use of vegetable oils in fish diets in the

1950s and 1960s might, in part, have been based on their greater stability in prepared diets.

It has been demonstrated that rancid herring and hake meals in fish feeds caused dark colouration, anaemia, lethargy, brown-yellow pigmented liver, abnormal kidneys, and small gill clubbing in chinook salmon.

The symptons can be alleviated by addition of alpha-tocopherol to the diets containing rancid fish meals. The addition of vitamin E would prevent the toxic or negative effects of adding 5 per cent highly oxidized salmon oil to the diet of rainbow trout. This same sparing effect of alpha-tocopherol can also apply to rancid carp feed.

The positive nutritional value of ω3 fatty acids in fish lipids for fish feeds can become a negative factor if adequate care is not taken in the preparation and storage of feeds. Only fresh oils with low peroxide values should be used in feeds. Fish feed ingredients such as fish meals should be protected against oxidation.

The level of vitamin E added to the diet should be increased as the PUFA level is increased. The finished feed, if possible, should be stored in air tight containers at reduced temperatures with minimum exposure to UV radiation and other factors accelerating the rate of lipid oxidation. The problems of rancidity or antioxidation of lipids in fish feeds should not be ignored.

Chapter 8

Insecticides

Insecticides are agents of chemical or biological origin that control insects. Control may result from killing the insect or otherwise preventing it from engaging in behaviours deemed destructive. Insecticides may be natural or manmade and are applied to target pests in a myriad of formulations and delivery systems (sprays, baits, slow-release diffusion, etc.). The science of biotechnology has, in recent years, even incorporated bacterial genes coding for insecticidal proteins into various crop plants that deal death to unsuspecting pests that feed on them.

The purpose is to provide a handshake overview of what insecticides are, and a short background and a review of the major insecticide classes that have been or are used today to cope with insect pests. Though by no means exhaustive, we will touch on major classes and technologies whether decades old or recently revealed.

Some 10,000 species of the more than 1 million species of insects are crop-eating, and of these, approximately 700 species worldwide cause most of the insect damage to man's crops, in the field and in storage. Humanoids have been on earth for more than 3 million years, while insects have existed for at least 250 million years. We can guess that among the first approaches used by our primitive ancestors to reduce insect annoyance was hugging smoky fires or spreading mud and dust over their skin to repel biting and tickling insects, a practice resembling the habits of elephants, swine, and water buffalo. Today, such approaches would be classed as *repellents*, a category of *insecticides.*

Historians have traced the use of pesticides to the time of Homer around 1000 B.C., but the earliest records of insecticides pertain to the burning of "brimstone" (sulfur) as a fumigant. Pliny the Elder recorded most of the earlier insecticide uses in his *Natural History*. Included among these were the use of gall from a green lizard to protect apples from worms and rot. Later, we find a variety of materials used with questionable results extracts of pepper and tobacco, soapy water, whitewash, vinegar, turpentine, fish oil, brine, lye among many others.

At the beginning of World War II, our insecticide selection was limited to several arsenicals, petroleum oils, nicotine, pyrethrum, rotenone, sulfur, hydrogen cyanide gas, and cryolite. It was World War II that opened the *Modern Era of Chemical* control with the introduction of a new concept of insect control –synthetic organic insecticides, the first of which was DDT.

ORGANOCHLORINES

The organochlorines are insecticides that contain carbon (thus *organo-*), hydrogen, and chlorine. They are also known by other names *chlorinated hydrocarbons, chlorinated organics, chlorinated insecticides*, and *chlorinated synthetics*. The organochlorines are now primarily of historic interest, since few survive in today's arsenal.

DIPHENYL ALIPHATICS

The oldest group of the organochlorines is the *diphenyl aliphatics*, which included DDT, DDD, dicofol, ethylan, chlorobenzilate, and methoxychlor. DDT is probably the best known and most notorious chemical of the 20th century.

It is also fascinating, and remains to be acknowledged as the most useful insecticide developed. More than 4 billion pounds of DDT were used throughout the world, beginning in 1940, and in the U.S. ending essentially in 1973, when the U.S. Environmental Protection Agency canceled all uses. The remaining First World countries rapidly followed suit. DDT is still effectively used for malaria control in several third world countries. In 1948, Dr. Paul Muller, a Swiss entomologist, was awarded the Nobel Prize in Medicine for his lifesaving discovery of DDT as an insecticide useful in the control of malaria, yellow fever and many other insect-vectored diseases.

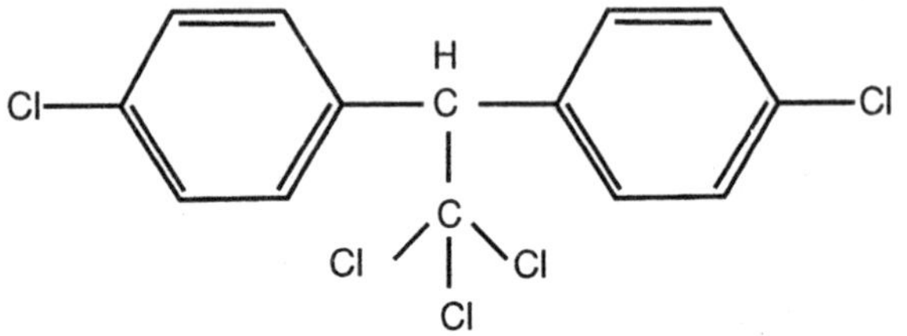

Mode of action–The mode of action for DDT has never been clearly established, but in some complex manner it destroys the delicate balance of sodium and potassium ions within the axons of the neuron in a way that prevents normal transmission of nerve impulses, both in insects and mammals.

It apparently acts on the sodium channel to cause "leakage" of sodium ions. Eventually the affected neurons fire impulses spontaneously, causing the muscles to twitch– "DDT jitters"– followed by convulsions and death. DDT has a negative temperature correlation–the lower the surrounding temperature the more toxic it becomes to insects.

Hexchlorocyclohexane (HCH)

Also known as benzenehexachloride (BHC), the insecticidal properties of HCH were discovered in 1940 by French and British entomologists. In its technical grade, there are five isomers, *alpha, beta, gamma, delta* and *epsilon*. Surprisingly, only the *gamma* isomer has insecticidal properties.

Consequently, the *gamma* isomer was isolated in manufacture and sold as the odorless insecticide *lindane*. In contrast, technical grade HCH has a strong musty odor and flavour, which can be imparted to treated crops and animal products. Because of its very low cost, HCH is still used in many developing countries. In 2002, the U.S. EPA removed all food-related (tolerance-requiring) uses of lindane from the U.S.

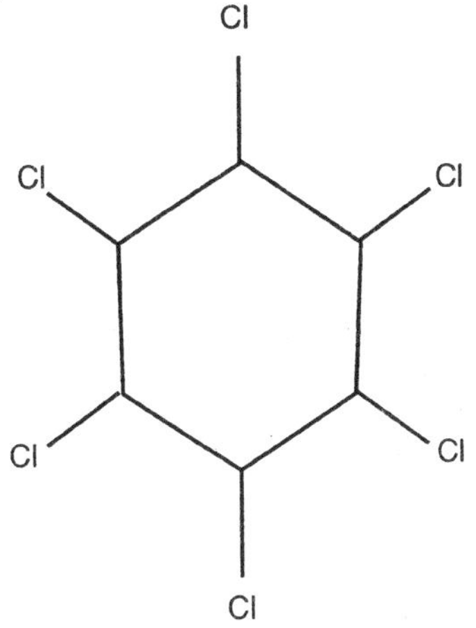

Mode of Action

The effects of HCH superficially resemble those of DDT, but occur much more rapidly, and result in a much higher rate of respiration in insects. The *gamma* isomer is a neurotoxicant whose effects are normally seen within hours as increased activity, tremors, and convulsions leading to prostration. It too, exhibits a negative temperature correlation, but not as pronounced as that of DDT.

Cyclodienes

The cyclodienes appeared after World War II chlordane, 1945, aldrin and dieldrin, 1948; heptachlor, 1949; endrin, 1951; mirex, 1954; endosulfan, 1956; and chlordecone, 1958. There were other cyclodienes of minor importance developed in the U.S. and Germany. Most of the cyclodienes are persistent insecticides and are stable in soil and relatively stable to the ultraviolet of sunlight. As a result, they were used in greatest quantity as soil insecticides (especially chlordane, heptachlor, aldrin, and dieldrin) for the control of termites and soil-borne insects whose larval stages feed on the roots of plants.

To appreciate the effectiveness of these materials as termiticides, consider that wood and wooden structures treated with chlordane, aldrin, and dieldrin in the year of their development are still protected from damage—after more than 60 years! The cyclodienes were the most effective, long-lasting and economical termiticides ever developed.

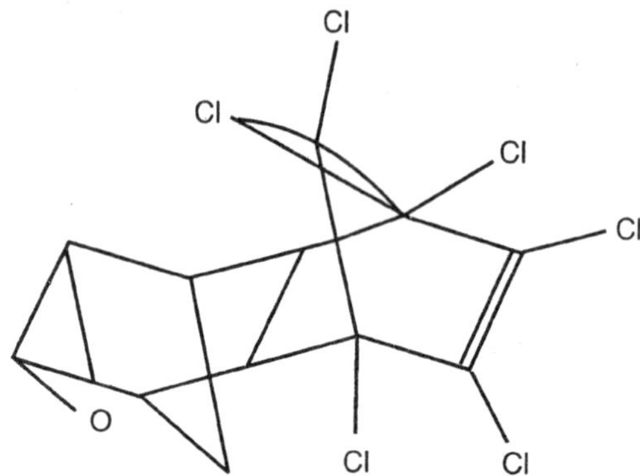

Because of their persistence in the environment, resistance that developed in several soil insect pests, and in some instances *biomagnification* in wildlife food chains, most agricultural uses of cyclodienes were canceled by the EPA between 1975 and 1980, and their use as termiticides canceled in 1984-88.

Mode of Action

Unlike DDT and HCH, the cyclodienes have a positive temperature correlation–their toxicity increases with increasing ambient temperature. Their modes of action are also not clearly understood. However, it is known that this group acts on the inhibitory mechanism called the GABA (g-aminobutyric acid) receptor.

This receptor operates by increasing chloride ion permeability of neurons. Cyclodienes prevent chloride ions from entering the neurons, and thereby antagonize the "calming" effects of GABA. Cyclodienes appear to affect all animals similarly, first with the nervous activity followed by tremors, convulsions and prostration.

Polychloroterpenes

Only two polychloroterpenes were developed–toxaphene in 1947, and strobane in 1951. Toxaphene had by far the greatest use of any single insecticide in agriculture, while strobane was relatively insignificant.

Toxaphene was used on cotton, first in combination with DDT, for alone it had minimal insecticidal qualities. Then, in 1965, after several major cotton insects became resistant to DDT, toxaphene was formulated with methyl parathion, an organophosphate insecticide mentioned later.

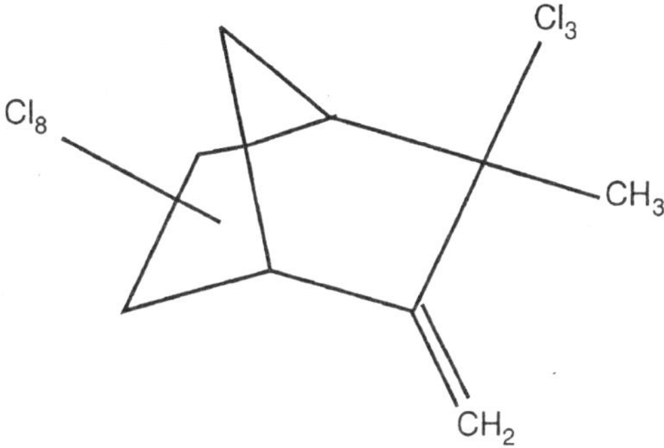

Toxaphene is a mixture of more than 177 10-carbon polychlorinated derivatives. These materials persist in the soil, though not as long as the cyclodienes, and disappear from the surfaces of plants in 3-4 weeks. This disappearance is attributed more to volatility than to photolysis or plant metabolism. Toxaphene is rather easily metabolized by mammals and birds, and is not stored in body fat nearly to the extent of DDT, HCH and the cyclodienes. Despite its low toxicity to insects, mammals and birds, fish are highly susceptible to toxaphene poisoning, in the same order of magnitude as to the cyclodienes. Toxaphene's registrations were canceled by EPA in 1983.

Mode of Action

Toxaphene and strobane act on the neurons, causing an imbalance in sodium and potassium ions, similar to that of the cyclodiene insecticides.

ORGANOPHOSPHATES

Organophosphates (OPs) is the term that includes all

insecticides containing phosphorus. Other names used, but no longer in vogue, are *organic phosphates, phosphorus insecticides, nerve gas relatives,* and *phosphoric acid esters.*

All organophosphates are derived from one of the phosphorus acids, and as a class are generally the most toxic of all pesticides to vertebrates. Because of the similarity of OP chemical structures to the "nerve gases," their modes of action are also similar. Their insecticidal qualities were first observed in Germany during World War II in the study of the extremely toxic OP nerve gases *sarin, soman,* and *tabun*. Initially, the discovery was made in search of substitutes for nicotine, which was heavily used as an insecticide but in short supply in Germany.

The OPs have two distinctive features they are generally more toxic to vertebrates than other classes of insecticides, and most are chemically unstable or nonpersistent.

It is this latter characteristic that brought them into agricultural use as substitutes for the persistent *organochorines*. Because of the relatively high toxicity of the OP's, EPA, under provisions of the Food Quality Protection Act (1996), undertook an extensive reappraisal of the entire class beginning in the late 1990's. Many OP's were voluntarily canceled and others lost uses.

Mode of action–The OPs work by inhibiting certain important enzymes of the nervous system, namely *cholinesterase* (ChE). The enzyme is said to be *phosphorylated* when it becomes attached to the phosphorous moiety of the insecticide, a binding that is irreversible. This inhibition results in the accumulation of acetylcholine (ACh) at the neuron/neuron and neuron/muscle (neuromuscular) junctions or synapses, causing rapid twitching of voluntary muscles and finally paralysis.

CLASSIFICATION

All OPs are esters of phosphorus having varying combinations of oxygen, carbon, sulfur and nitrogen attached, resulting in six different subclasses phosphates, phosphonates, phosphorothioates, phosphorodithioates, phosphorothiolates and phosphoramidates. These subclasses are easily identified by their chemical names. The OPs are generally divided into three groups–*aliphatic, phenyl,* and *heterocyclic* derivatives.

ALIPHATICS

The aliphatic OPs are carbon chain-like in structure. The first OP brought to agriculture, TEPP belonged to this group. Other examples are malathion, trichlorfon, monocrotophos, dimethoate, oxydemetonmethyl, dicrotophos, disulfoton, dichlorvos, mevinphos, methamidophos, and acephate.

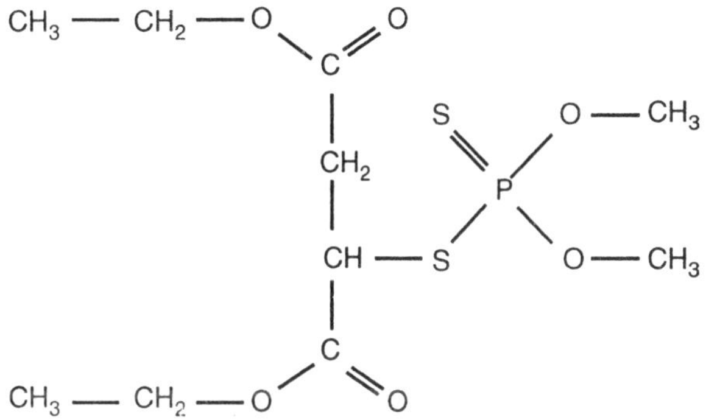

CH_3 — CH_2 — O   O
C
S   O — CH_3
CH_2
P
CH — S   O — CH_3
C
CH_3 — CH_2 — O   O

PHENYL DERIVATIVES

The phenyl OPs contain a phenyl ring with one of the ring hydrogens displaced by attachment to the phosphorus moiety and other hydrogens frequently displaced by Cl, NO_2, CH_3, CN, or S. The phenyl OPs are generally more stable than the aliphatics, thus their residues are longer lasting. The first phenyl OP brought into agriculture was parathion in 1947. Examples of other phenyl OPs are methyl parathion, profenofos, sulprofos, isofenphos, fenitrothion, fenthion, and famphur.

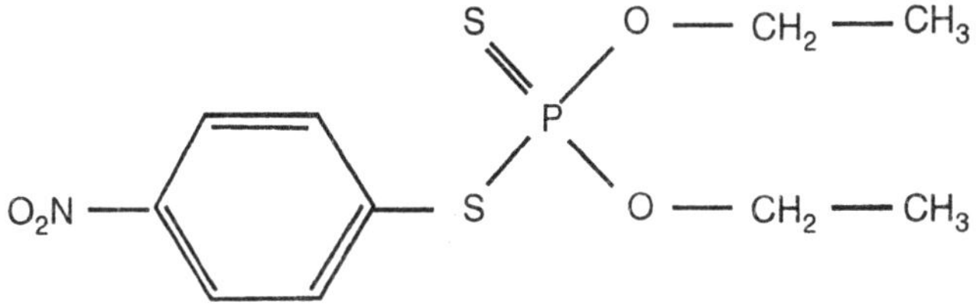

S   O — CH_2 — CH_3
P
O_2N — S   O — CH_2 — CH_3

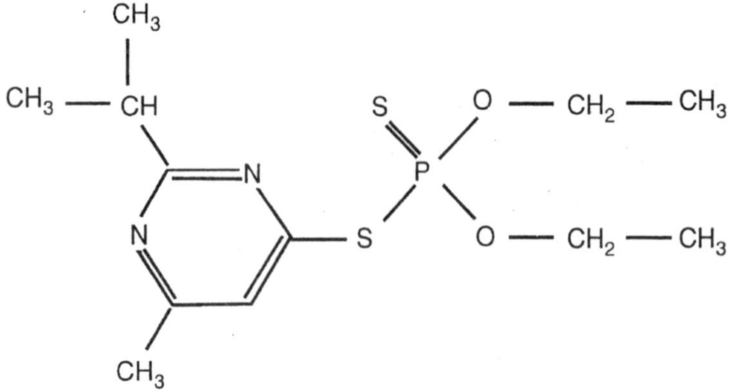

Heterocyclic derivatives–The term *heterocyclic* means that the ring structures are composed of different or unlike atoms, e.g.,oxygen, nitrogen or sulfur. The first of this group was diazinon introduced in 1952. Other examples in this group are azinphos-methyl, azinphos-ethyl, chlorpyrifos, methidathion, phosmet, isazophos, and chlorpyrifos-methyl.

ORGANOSULFURS

These few materials have very low toxicity to insects and are used only as acaricides (miticides). They contain two phenyl rings, resembling DDT, with sulfur in place of carbon as the central atom. These include tetradifon, propargite, and ovex.

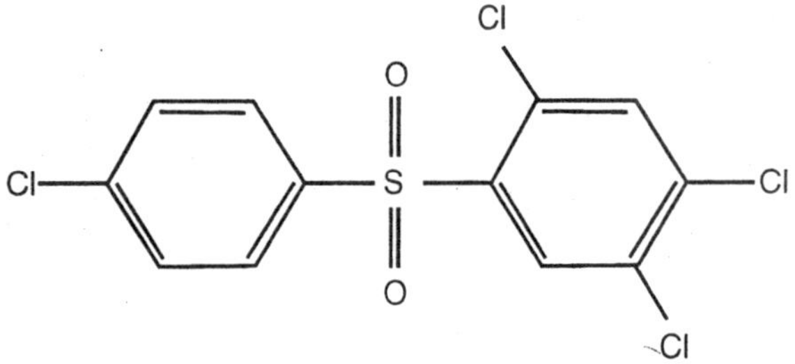

CARBAMATES

The carbamate insecticides are derivatives of carbamic acid (as the OPs are derivatives of phosphoric acid). And like the OPs, their mode of action is that of inhibiting the vital enzyme *cholinesterase* (ChE).

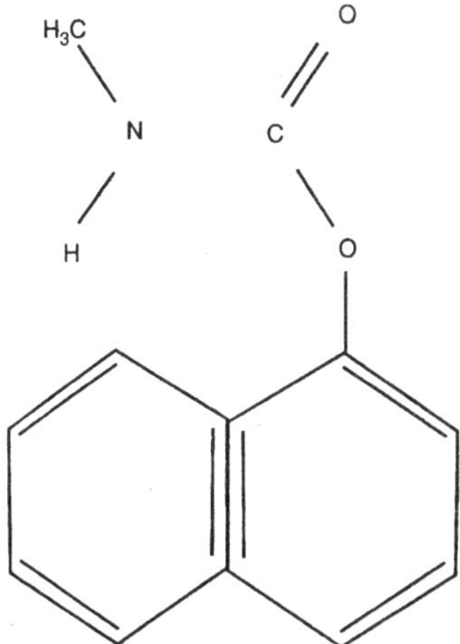

The first successful carbamate insecticide, carbaryl, was introduced in 1956. More of it has been used worldwide than all the remaining carbamates combined. Two distinct qualities have made it the most popular carbamate its very low mammalian oral and dermal toxicity and an exceptionally broad spectrum of insect control. Other long-standing carbamate insecticides are methomyl, carbofuran, aldicarb, oxamyl, thiodicarb, methiocarb, propoxur, bendiocarb, carbosulfan, aldoxycarb, promecarb, and fenoxycarb. Carbamates more recently introduced include primicarb, indoxacarb, alanycarb and furathiocarb.

Mode of Action

Carbamates inhibit cholinesterase (ChE) as OPs do, and they behave in almost identical manner in biological systems, but with two main differences. Some carbamates are potent inhibitors of aliesterase (miscellaneous aliphatic esterases whose exact functions are not known), and their selectivity is sometimes more pronounced against the ChE of different species. Second, ChE inhibition by carbamates is reversible. When ChE is inhibited by a carbamate, it is said to be *carbamylated,* as when an OP results in the enzyme being *phosphorylated.* In insects, the effects of OPs and carbamates are primarily those of poisoning of the central nervous system, since the insect neuromuscular junction is not cholinergic, as in mammals. The only cholinergic synapses known in insects are in the central nervous system. (The chemical neuromuscular junction transmitter in insects is thought to be glutamic acid.)

FORMAMIDINES

The formamidines comprise a small group of insecticides. Three examples are chlordimeform, which is no longer registered in the U.S., formetanate, and amitraz. Their current value lies in the control of OP- and carbamate-resistant pests.

Mode of action–Formamidine poisoning symptoms are distinctly different from other insecticides.

Their proposed action is the inhibition of the enzyme monoamine oxidase, which is responsible for degrading the neurotransmitters norepinephrine and serotonin. This results in the accumulation of these compounds, which are known as *biogenic amines.* Affected insects become quiescent and die.

DINITROPHENOLS

The basic dinitrophenol molecule has a broad range of toxicities–as herbicides, insecticides, ovicides, and fungicides. Of the insecticides, binapacryl and dinocap were the most recently used. Dinocap is an effective miticide and was very heavily used as a fungicide for the control of powdery mildew fungi. Because of the inherent toxicity of the dinitrophenols, they have all been withdrawn.

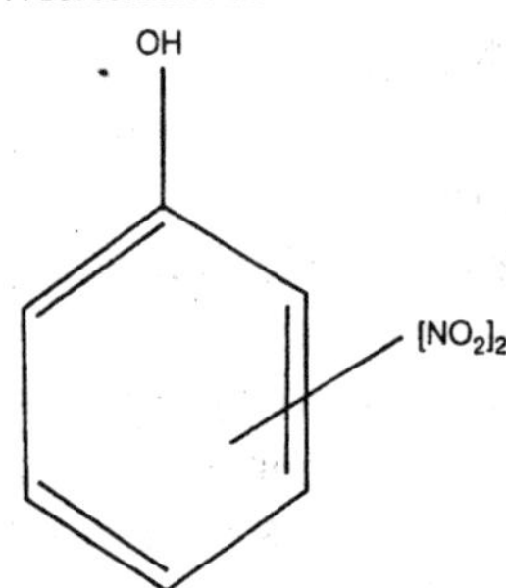

Mode of action–Dinitrophenols act by uncoupling or inhibiting oxidative phosphorylation, which basically prevents the formation of the high-energy phosphate molecule, adenosine triphosphate (ATP).

ORGANOTINS

The organotins are a group of acaricides that double as fungicides. Of particular interest is cyhexatin, one of the most selective acaricides known, introduced in 1967. Fenbutatin-oxide has been used extensively against mites on deciduous fruits, citrus, greenhouse crops, and ornamentals.

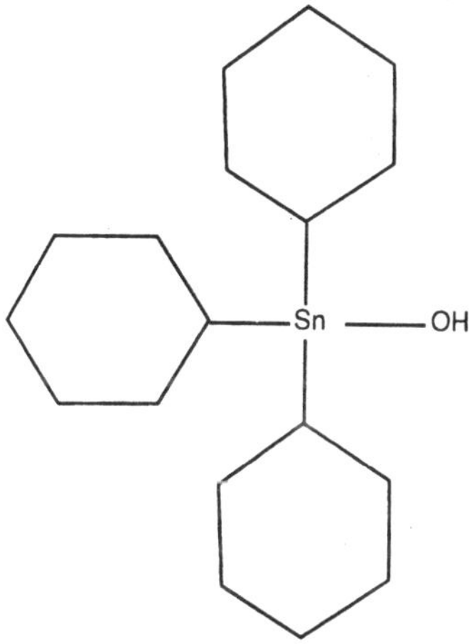

Sn —— OH

Mode of action–These tin compounds inhibit oxidative phosphorylation at the site of dinitrophenol uncoupling, preventing the formation of the high-energy phosphate molecule adenosine triphosphate (ATP). These trialkyl tins also inhibit photophosphorylation in chloroplasts, the chlorophyll-bearing subcellular units) and could therefore serve as algicides.

PYRETHROIDS

Natural pyrethrum has seldom been used for agricultural purposes because of its cost and instability in sunlight. In recent decades, many synthetic pyrethrin-like materials have become available. They were originally referred to as *synthetic pyrethroids*. Currently, the better nomenclature is simply *pyrethroids*. These are stable in sunlight and are generally effective against most agricultural insect pests when used at the very low rates of 0.01 to 0.1 pound per acre.

The pyrethroids have an interesting evolution, which is conveniently divided into four generations. The first generation contains only one pyrethroid, allethrin, which appeared in 1949. Its synthesis was very complex, involving 22 chemical reactions to reach the final product.

The second generation includes tetramethrin, followed by resmethrin in 1967 (20X as effective as pyrethrum), then bioresmethrin, then Bioallethrin, and finally phonothrin. The third generation includes fenvalerate, and permethrin which appeared in 1972-73. These became the first agricultural pyrethroids because of their exceptional insecticidal activity and their photostability. They were virtually unaffected by ultraviolet in sunlight, lasting 4-7 days as efficacious residues on crop foliage.

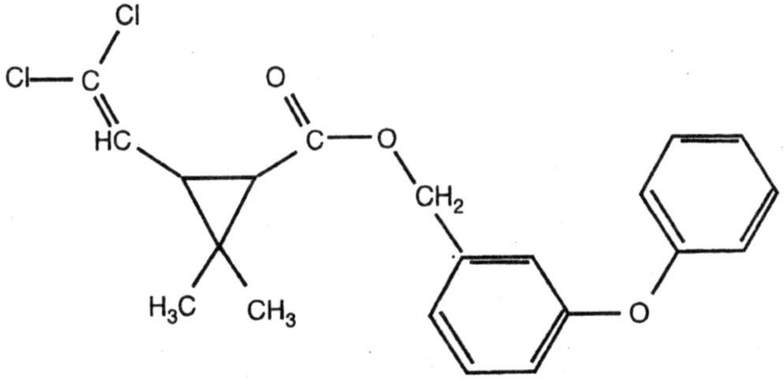

Cl
Cl—C
HC
O
C—O
CH_2
H_3C CH_3
O

The fourth and current generation, is truly exciting because of their effectiveness in the range of 0.01 to 0.05 lb ai/A. These include bifenthrin, *lambda*-cyhalothrin, cypermethrin, cyfluthrin, deltamethrin, fenpropathrin, flucythrinate, fluvalinate, prallethrin, *tau*-fluvalinate tefluthrin, tralomethrin, and *zeta*-cypermethrin.

All of these are photostable, that is, they do not undergo photolysis (splitting) in sunlight. And because they have minimal volatility they provide extended residual effectiveness, up to 10 days under optimum conditions.

Recent additions to the fourth generation pyrethroids are acrinathrin, imiprothrin, registered in 1998, and *gamma*-cyhalothrim, which is in development.

Mode of Action

The pyrethroids share similar modes of action, resembling that of DDT, and are considered axonic poisons. They apparently work by keeping open the sodium channels in neuronal membranes. There are two types of pyrethroids.

Type I, among other physiological responses, have a negative temperature coefficient, resembling that of DDT. Type II, in contrast have a positive temperature coefficient, showing increased kill with increase in ambient temperature.

Pyrethroids affect both the peripheral and central nervous system of the insect. They initially stimulate nerve cells to produce repetitive discharges and eventually cause paralysis. Such effects are caused by their action on the sodium channel, a tiny hole through which sodium ions are permitted to enter the axon to cause excitation. The stimulating effect of pyrethroids is much more pronounced than that of DDT.

NICOTINOIDS

The nicotinoids are a newer class of insecticides with a new mode of action. They have been previously referred to as *nitro-quanidines, neonicotinyls, neonicotinoids, chloronicotines,* and more recently as the *chloronicotinyls.*

Just as the synthetic pyrethroids are similar to and modeled after the natural pyrethrins, so too, are the nicotinoids similar to and modeled after the natural nicotine ÷ Imidacloprid was introduced in Europe and Japan in 1990 and first registered in the U.S. in 1992. It is currently marketed as several proprietary products worldwide, e.g., Admire, Confidor, Gaucho, Merit, Premier, Premise and Provado. Very possibly it is used in the greatest volume globally of all insecticides.

Imidacloprid is a systemic insecticide, having good root-systemic characteristics and notable contact and stomach action. It is used as a soil, seed or foliar treatment in cotton, rice cereals, peanuts, potatoes, vegetables, pome fruits, pecans and turf, for the control of sucking insects, soil insects, whiteflies, termites, turf insects and the Colourado potato beetle, with long residual control. Imidacloprid has no effect on mites or nematodes.

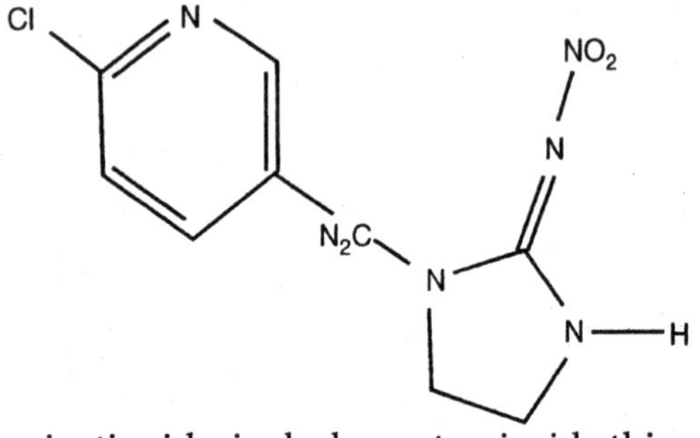

Cl N NO_2 N N_2C N N—H

Other nicotinoids include acetamiprid, thiamethoxam, nitenpyram, clothianidin, dinotefuran and thiacloprid. U.S. registrations for acetamiprid, thiamethoxam and thiacloprid were granted in 2002 and clothianidin in 2003.

Mode of Action

The nicotinoids act on the central nervous system of insects, causing irreversible blockage of postsynaptic nicotinergic acetylcholine receptors.

SPINOSYNS

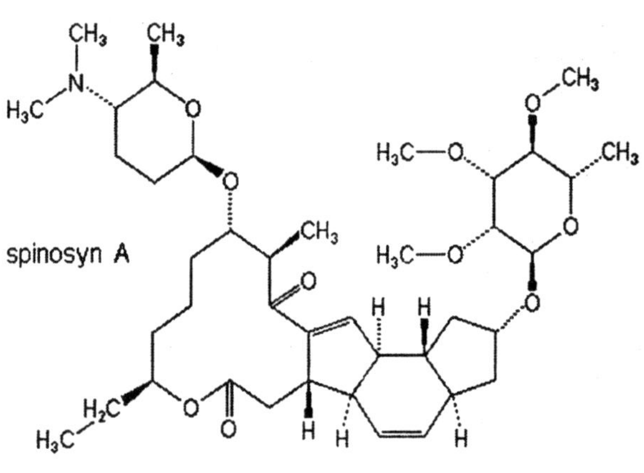

CH_3 CH_3 N H_3C O O CH_3 O H_3C—O CH_3 O CH_3 spinosyn A H_3C—O O O O H H H_2C H_3C O O H H H

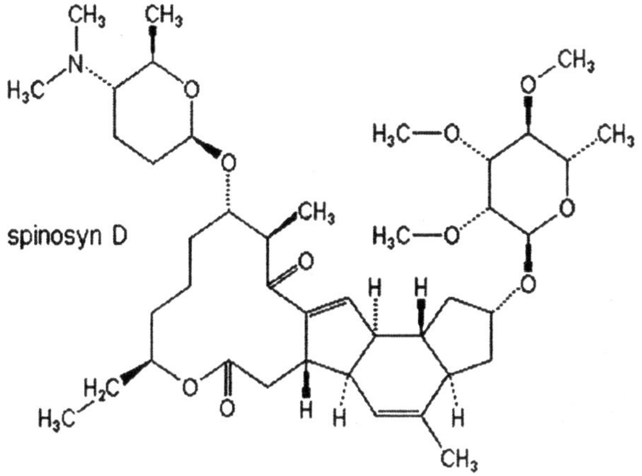

spinosyn D

Spinosyns are among the newest classes of insecticides, represented by spinosad. Spinosad is a fermentation metabolite of the actinomycete *Saccharopolyspora spinosa,* a soil-inhabiting microorganism.

It has a novel molecular structure and mode of action that provide excellent crop protection typically associated with synthetic insecticides, first registered for use on cotton in 1997. Spinosad is a mixture of spinosyns A and D. It is particularly effective as a broad-spectrum material for most caterpillar pests at the astonishing rates of 0.04 to 0.09 pound of active ingredient (18 to 40 grams) per acre. It has both contact and stomach activity against lepidopteran larvae, leaf miners, thrips, and termites, with long residual activity. Crops registered include cotton, vegetables, tree fruits, ornamentals and others.

Mode of action–Spinosad acts by disrupting binding of acetylcholine in nicotinic acetylcholine receptors at the postsynaptic cell.

FIPROLES

Fipronil is the only insecticide in this new class, introduced in 1990 and registered in the U.S. in 1996. It is a systemic material with contact and stomach activity. Fipronil is used for the control of many soil and foliar insects, (e.g., corn rootworm, Colourado potato beetle, and rice water weevil) on a variety of crops, primarily corn, turf, and for public health insect control. It is also used for seed treatment and formulated as baits for cockroaches, ants and termites. Fipronil is effective against insects resistant or tolerant to pyrethroid, organophosphate and carbamate insecticides.

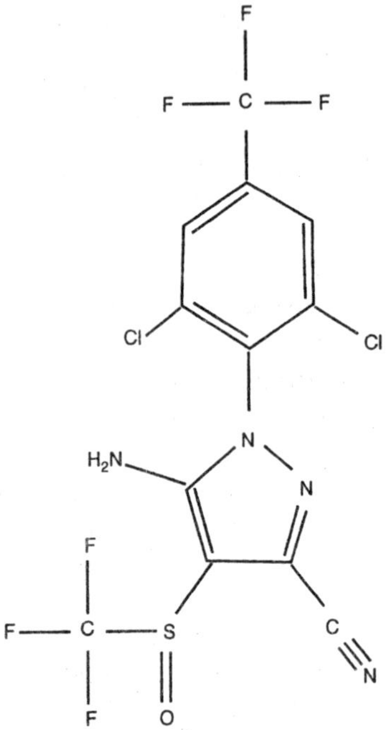

Mode of Action

Fipronil blocks the (g-aminobutyric acid- (GABA) regulated chloride channel in neurons, thus antagonizing the "calming" effects of GABA, similar to the action of the Cyclodienes.

PYRROLES

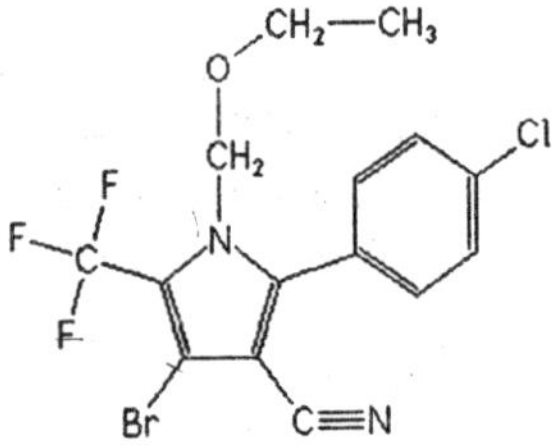

Chlorfenapyr is the first and only member of this unique chemical group, as both a contact and stomach insecticide-miticide. It is used on cotton and experimentally on corn, soybeans, vegetables, tree and vine crops, and ornamentals to control whitefly, thrips, caterpillars, mites, leafminers, aphids, and Colourado potato beetle. It has ovicidal activity on some species. EPA took the unusual step of refusing to register chlorfenapyr in 2000 for cotton insect control because of potential hazards to birds. However, labels for greenhouse ornamentals were granted in 2001.

Mode of Action

Chlorfenapyr is an "uncoupler" or inhibitor of oxidative phosphorylation, preventing the formation of the crucial energy molecule adenosine triphosphate (ATP).

PYRAZOLES

The original pyrazoles were tebufenpyrad and fenpyroximate (not illustrated). These were designed primarily as non-systemic contact and stomach miticides, but do have limited effectiveness on psylla, aphids, whitefly, and thrips. Tebufenpyrad, registered by EPA in 2002, is used on cotton, soybeans, vegetables, pome fruits, grapes and citrus. Fenpyroximate controls all stages of mites, gives fast knockdown, inhibits molting of immature stages of mites, and has long residual activity. Newer members of this class include ethiprole which is active on a broad sprectum of chewing and sucking insects, and tolfenpyrad which is reputed to active on pests infesting cole and cucurbit crops.

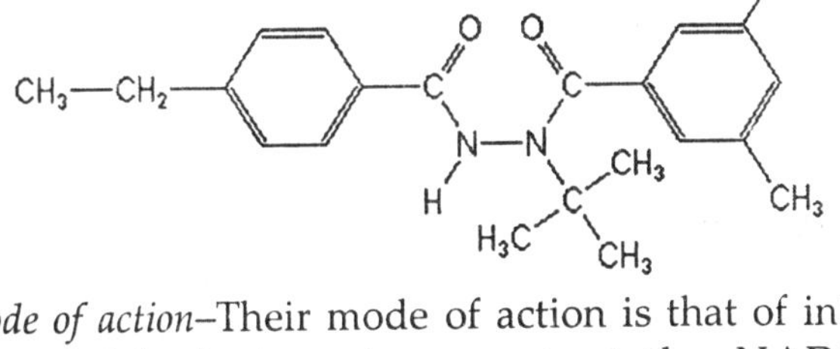

CH_3—CH_2—(benzene ring)—C(=O)—N(H)—N(C(CH_3)(CH_3)(H_3C))—C(=O)—(benzene ring with two CH_3)

Mode of action–Their mode of action is that of inhibiting mitochondrial electron transport at the NADH-CoQ reductase site, leading to the disruption of adenosine triphosphate (ATP) formation, the crucial energy molecule.

PYRIDAZINONES

Pyridaben is the only member of this class. It is a selective contact insecticide and miticide, also effective against thrips, aphids, whiteflies and leafhopprs. Registrations are for pome fruits, almonds, citrus, ornamentals and greenhouse ornamentals. Pyridaben provides exceptionally long residual control, and rapid knockdown at a broad range of temperatures.

Mode of Action

Pyridaben is a metabolic inhibitor that interrupts mitochondrial electron transport at Site 1, similar to the Quinazolines, below.

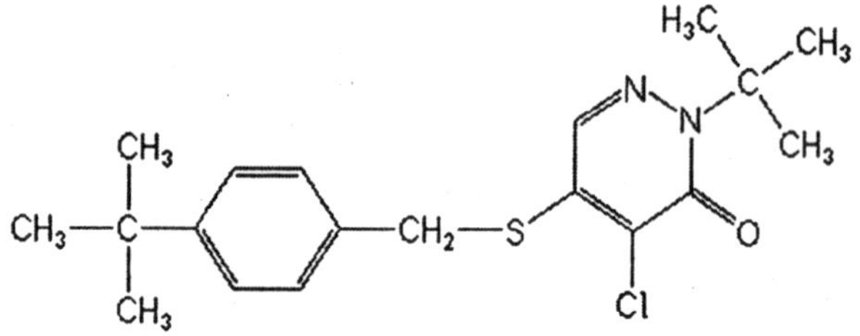

H_3C CH_3 CH_3 CH_3 CH_3—C— —CH_2—S N N C O Cl CH_3

QUINAZOLINES

The quinazolines offer a unique chemical configuration, consisting only of one insecticide, fenazaquin. Fenazaquin is a contact and stomach miticide. It has ovicidal activity, gives rapid knockdown, and controls all stages of mites. Not yet registered in the U.S., it is used on cotton, stone and pome fruits, citrus, grapes and ornamentals.

FENAZAQUIN (MATADOR)

Mode of action–Fenazaquin inhibits mitochondrial electron transport at Site 1, similar to the Pyridazinones, above.

BENZOYLUREAS

Benzoylureas are an entirely different class of insecticides that act as insect growth regulators (IGRs). Rather than being the typical poisons that attack the insect nervous system, they interfere with chitin synthesis and are taken up more by ingestion than by contact. Their greatest value is in the control of caterpillars and beetle larvae.

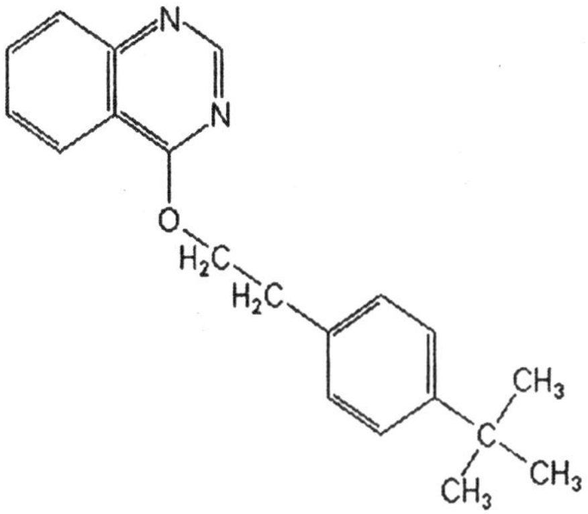

Benzoylureas were first used in Central America in 1985, to control a severe, resistant leafworm complex (*Spodoptera, Trichoplusia*) outbreak in cotton. The withdrawal of the ovicide chlordimeform made their control quite difficult due to their high resistance to almost all insecticide classes, including the pyrethroids. The benzoylureas were introduced in 1978 by Bayer of Germany, triflumuron being the first. Others appearing since then are chlorfluazuron, followed by teflubenzuron, hexaflumuron, flufenoxuron, and flucycloxuron. Others are flurazuron, novaluron, and diafenthiuron, bistrifluron and noviflumuron. Until recently lufenuron was the newest addition to this group, appearing in 1990. Among the newer benzoylureas only hexaflumuron and novaluron have been registered by EPA.

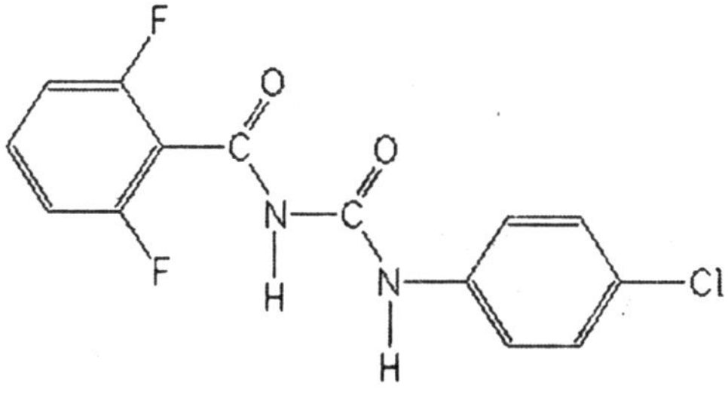

The only other benzoylurea registered in the U.S. is diflubenzuron. It was first registered in 1982 for gypsy moth, cotton boll weevil, most forest caterpillars, soybean caterpillars, and mushroom flies, but now with a much broader range of registrations. Though not a benzoylurea, cyromazine, a triazine, is also a potent chitin synthesis inhibitor. It is selective toward Dipterous species and used for the control of leafminers in vegetable crops and ornamentals, and fed to poultry or sprayed to control flies in manure of broiler and egg producing operations, and incorporated into compost of mushroom houses for fungus gnats.

Mode of Action

The benzoylureas act on the larval stages of most insects by inhibiting or blocking the synthesis of chitin, a vital and almost indestructible part of the insect exoskeleton. Typical effects on developing larvae are the rupture of malformed cuticle or death by starvation. Adult female boll weevils exposed to diflubenzuron lay eggs that do not hatch. And, mosquito larvae control can be achieved with as little as 1.0 gram of diflubenzuron per acre of surface water.

BOTANICALS

Botanical insecticides are of great interest to many, for they are *natural* insecticides, toxicants derived from plants. Historically, the plant materials have been in use longer than any other group, with the possible exception of sulfur. Tobacco, pyrethrum, derris, hellebore, quassia, camphor, and turpentine were some of the more important plant products in use before the organized search for insecticides began in the early 1940s.

In recent years the term *biorational* has been put into play by the EPA. There are similarities and differences between the terms botanical and biorational. Botanical insecticide use in the U.S. peaked in 1966, and has declined steadily since. Pyrethrum is now the only classical botanical of significance in use. Some newer plant-derived insecticides that have come into use are referred to as *florals or scented plant chemicals* and include, among others, limonene, cinnamaldehyde and eugenol. In addition, there is azadirachtin from the neem tree which is used in greenhouse and on ornamentals.

PYRETHRUM

Pyrethrum is extracted from the flowers of a chrysanthemum grown in Kenya and Ecuador. It is one of the oldest and safest insecticides available. The ground, dried flowers were used in the early 19th century as the original louse powder to control body lice in the Napoleonic Wars.

Pyrethrum acts on insects with phenomenal speed causing immediate paralysis, thus its popularity in fast knockdown household aerosols. However, unless it is formulated with one of the *synergists*, most of the paralyzed insects recover to once again become pests. Pyrethrum is a mixture of four compounds pyrethrins I and II and cinerins I and II.

Mode of action–Pyrethrum is an axonic poison, as are the synthetic pyrethroids and DDT. Axonic poisons are those that in some way affect the electrical impulse transmission along the axons, the elongated extensions of the neuron cell body. Pyrethrum and some pyrethroids have a greater insecticidal effect when the temperature is lowered, a negative temperature coefficient, as does DDT.

They affect both the peripheral and central nervous system of the insect. Pyrethrum initially stimulates nerve cells to produce repetitive discharges, leading eventually to paralysis. Such effects are caused by their action on the sodium channel, a tiny hole through which sodium ions are permitted to enter the axon to cause excitation. These effects are produced in insect nerve cord, which contains ganglia and synapses, as well as in giant nerve fibre axons.

NICOTINE

Nicotine is extracted by several methods from tobacco,

and is effective against most all types of insect pests, but is used particularly for aphids and caterpillars–soft bodied insects.

Nicotine is an alkaloid, a chemical class of heterocyclic compounds containing nitrogen and having prominent physiological properties. Other well-known alkaloids that are not insecticides are caffeine (coffee, tea), quinine (cinchona bark), morphine (opium poppy), cocaine (coca leaves), ricinine (a poison in castor oil beans), strychnine, coniine (spotted hemlock, the poison used by Socrates), and, finally LSD (a hallucinogen from the ergot fungus attacking grain).

Mode of Action

Nicotine action is one of the first, classic modes of action identified by pharmacologists. Drugs that act similarly to nicotine are said to have a nicotinic response. Nicotine mimics acetylcholine (ACh) at the neuromuscular (nerve/muscle) junction in mammals, and results in twitching, convulsions, and death, all in rapid order. In insects the same action is observed, but only in the central nervous system ganglia.

Rotenone

Rotenone or rotenoids are produced in the roots of two genera of the legume family *Derris* and *Lonchocarpus* (also called cubé) grown in South America. It is both a stomach and contact insecticide and used for the last century and a half to control leaf-eating caterpillars, and three centuries prior to that in South America to paralyze fish, causing them to surface and be easily captured. Today, rotenone is used in the same way to reclaim lakes for game fishing. Used on a prescribed basis, it eliminates all fish, closing the lake to reintroduction of rough species. It is a selective piscicide in that it kills all fish at dosages that are relatively nontoxic to fish food organisms, and is degraded rapidly.

Mode of Action

Rotenone is a respiratory enzyme inhibitor, acting between NAD+ (a coenzyme involved in oxidation and reduction in metabolic pathways) and coenzyme Q (a respiratory enzyme responsible for carrying electrons in some electron transport chains), resulting in failure of the respiratory functions.

LIMONENE OR D-LIMONENE

Limonene or *d*-Limonene is the latest addition to the botanicals. Limonene belongs to a group often called *florals* or *scented plant chemicals*. Extracted from citrus peel, it is effective against all external pests of pets, including fleas, lice, mites, and ticks, and is virtually nontoxic to warm-blooded animals.

Several insecticidal substances occur in citrus oil, but the most important is limonene, which constitutes about 98% of the orange peel oil by weight.

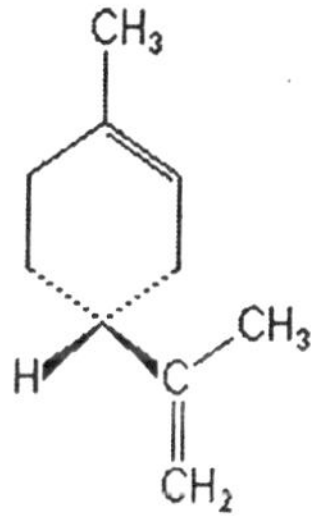

CH_3

H C CH_3

CH_2

Two other recently introduced floral products are eugenol (Oil of Cloves) and cinnamaldehyde (derived from Ceylon and Chinese cinnamon oils). They are used on ornamentals and many crops to control various insects.

Mode of Action

Its mode of action is similar to that of pyrethrum. It affects the sensory nerves of the peripheral nervous system, but it is not a ChE inhibitor.

Neem

Neem oil extracts are squeezed from the seeds of the neem tree and contain the active ingredient *azadirachtin*, a nortriterpenoid belonging to the lemonoids. Azadirachtin has shown some rather sensational insecticidal, fungicidal and bactericidal properties, including insect growth regulating qualities. Azatin is marketed as an insect growth regulator, and Align–and Nemix as a stomach/contact insecticide for greenhouse and ornamentals.

Mode of Action

Azadirachtin disrupts molting by inhibiting biosynthesis or metabolism of ecdysone, the juvenile molting hormone.

Synergists or Activators

Synergists are not in themselves considered toxic or insecticidal, but are materials used with insecticides to synergize or enhance the activity of the insecticides. The first was introduced in 1940 to increase the effectiveness of pyrethrum. Since then many materials have appeared, but only a few are still marketed. Synergists are found in most all household, livestock and pet aerosols to enhance the action of the fast knockdown insecticides pyrethrum, allethrin, and resmethrin, against flying insects. Current synergists, such as piperonyl butoxide, contain the methylenedioxyphenyl moiety, a molecule found in sesame oil and later named *sesamin*.

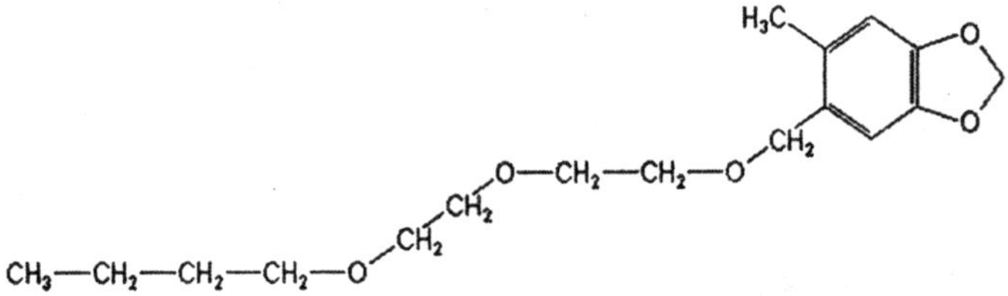

Mode of action–The synergists inhibit cytochrome P-450 dependent polysubstrate monooxygenases (PSMOs), enzymes produced by microsomes, the subcellular units found in the liver of mammals and in some insect tissues (e.g., fat bodies). The earlier name for these enzymes was mixed-function oxidases (MFOs). These PSMOs bind the enzymes that degrade selected foreign substances, such as pyrethrum, allethrin, resmethrin or any other synergized compound. Synergists simply bind the oxidative enzymes and prevent them from degrading the toxicant.

ANTIBIOTICS

In this category belong the *avermectins,* which are insecticidal, acaricidal, and antihelminthic agents that have been isolated from the fermentation products of *Streptomyces avermitilis,* a member of the actinomycete family.

Abamectin is the common name assigned to the avermectins, a mixture of containing 80% avermectin B1*a* and 20% B1*b*, homologs that have about equal biological activity. Clinch is a fire ant bait, and Avid is applied as a miticide/ insecticide. Abamectin has certain local systemic qualities, permitting it to kill mites on a leaf's underside when only the upper surface is treated.

The most promising uses for these materials are the control of spider mites, leafminers and other difficult-to-control greenhouse pests, and internal parasites of domestic animals. Emamectin benzoate is an analog of abamectin, produced by the same fermentation system as abamectin.

It was first registered in 1999. It is both a stomach and contact insecticide used primarily for control of caterpillars at the rate of 0.0075 to 0.015 lb (3.5 to 7.0 grams) a.i. per acre. Shortly after exposure, larvae stop feeding and become irreversibly paralyzed, dying in 3-4 days. Rapid photodegradation of both abamectin and emamectin occurs on the leaf surface.

More recently, milabectin has been introduced. It is a miticide with activity on piercing/sucking insects and is pending for registration.

Mode of action–Avermectins block the neurotransmitter (g-aminobutyric acid (GABA) at the neuromuscular junction in insects and mites. Visible activity, such as feeding and egg laying, stops shortly after exposure, though death may not occur for several days.

FUMIGANTS

The fumigants are small, volatile, organic molecules that become gases at temperatures above 40°F. They are usually heavier than air and commonly contain one or more of the halogens (Cl, Br, or F). Most are highly penetrating, reaching through large masses of material. They are used to kill insects, insect eggs, nematodes, and certain microorganisms in buildings, warehouses, grain elevators, soils, and greenhouses and in packaged products such as dried fruits, beans, grain, and breakfast cereals.

Although its use is now in decline because of environmental concerns, methyl bromide is the most heavily used of the fumigants, 68,424 metric tons worldwide in 1996, almost half of which is used in the U.S. The dominant use is for preplanting soil treatments, which accounted for 70% of that global total.

Quarantine uses account for 5-8%, while 8% is used to treat perishable products, such as flowers and fruits, and 12% for nonperishable products, like nuts and timber. Approximately 6% is used for structural applications, as in drywood termite fumigation of infested buildings.

With the recently passed change to the Clean Air Act amendments of 1990, U.S. production and importation must be reduced 25% from 1991 levels by 1999. A 50% reduction must be achieved by 2001, followed by a 70% reduction in 2003, and a full ban of the product in 2005. Under the Montreal Protocol, developing countries have until 2015 to phase out methyl bromide production. Some of the other common fumigants are ethylene dichloride, hydrogen cyanide, sulfuryl fluoride, Vapam, TeloneII, D-D, chlorothene, ethylene oxide, and the familiar home-use moth repellents napthalene crystals and paradichlorobenzene crystals.

Phosphine gas (PH_3) has also replaced methyl bromide in a few applications, primarily for insect pests of grain and food commodities. Treatment requires the use of aluminum or magnesium phosphide pellets, which react with atmospheric moisture to produce the gas. Phosphine, however, is very damaging to fresh commodities and is highly adsorbed onto oil, thus does not perform as a soil fumigant.

Alternatives that can fully replace methyl bromide are unlikely to be available by the deadlines set for replacement. Its low cost and utility on a wide variety of pests are hard to match. Because the loss of MeBr has considerable economic consequences, EPA has made it a priority to find and register replacements. To this end some progress has been made.

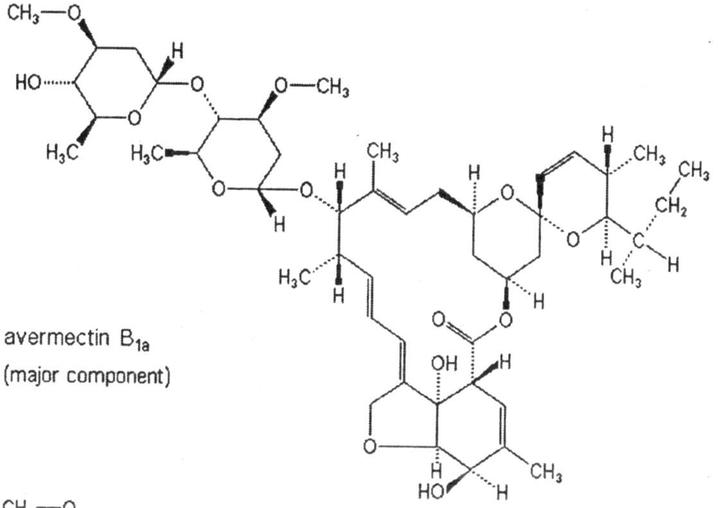

avermectin B_{1a}
(major component)

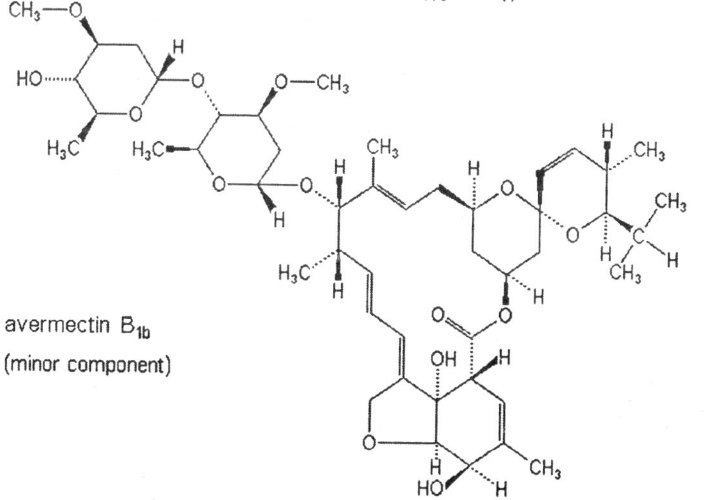

avermectin B_{1b}
(minor component)

The chemical 1,3-dichloropropane was registered in 2001 for preplant soil fumigation in strawberries and tomatoes. Moreover, iodomethane and metam-potassium are both being evaluated as soil fumigants.

Mode of action–Fumigants, as a group, are narcotics. That is, they act through means more physical than physiological.

The fumigants are liposoluble (fat soluble); they have common symptomology; their effects are reversible; and their activity is altered very little by structural changes in their molecules. As narcotics, they induce narcosis, sleep, or unconsciousness, which in effect is their action on insects.

Liposolubility appears to be an important factor in their action, since these narcotics lodge in lipid-containing tissues found throughout the insect body, including their nervous system.

INSECT REPELLENTS

Historically, repellents have included smoke, plants hung in dwellings or rubbed on the skin as the fresh plant or its brews, oils, pitches, tars, and various earths applied to the body. Before a more edified approach to insect olfaction and behaviour was developed, it was wrongly assumed that if a substance was repugnant to humans it would likewise be repellent to annoying insects.

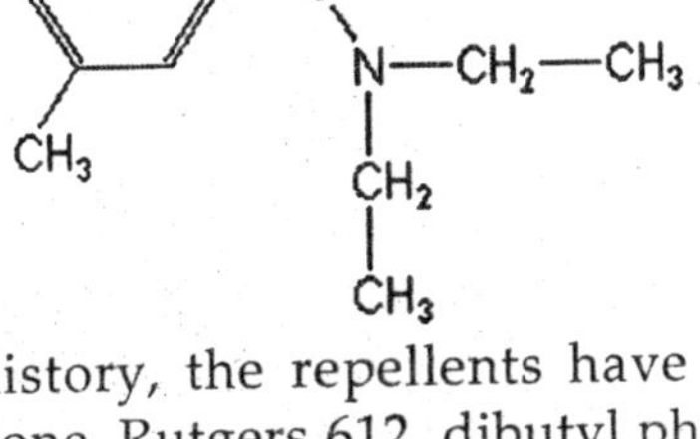

O
C
N—CH_2—CH_3
CH_3
CH_2
CH_3

In recent history, the repellents have been dimethyl phthalate, Indalone, Rutgers 612, dibutyl phthalate, various MGK repellents, benzyl benzoate, the military clothing repellent (N-butyl acetanilide), dimethyl carbate and diethyl toluamide.

Of these, only DEET has survived, and is used worldwide for biting flies and mosquitoes. Most of the others have lost their registrations and are no longer available. In 1999, EPA has registered a new insect repellent, N-methylneodecanamide. Rather than being used on humans to repel insects, it is applied to household floors and other surfaces to repel cockroaches and ants.

INORGANICS

Inorganic insecticides are those that do not contain carbon. Usually they are white crystals in their natural state, resembling the salts. They are stable chemicals, do not evapourate, and are usually water soluble. Sulfur, mentioned in the introduction, is very likely the oldest known, effective insecticide.

Sulfur and sulfur candles were burned by our great-grandparents for every conceivable purpose, from bedbug fumigation to the cleansing of a house just removed from quarantine of smallpox. Today, sulfur is a highly useful material in integrated pest management programmes where target pests specificity is important.

Sulfur dusts are especially toxic to mites of every variety, such as chiggers and spider mites, and to thrips and newly-hatched scale insects. Sulfur dusts and sprays are also fungicidal, particularly against powdery mildews. Several other inorganic compounds have been used as insecticides mercury, boron, thallium, arsenic, antimony, selenium, and fluoride.

Arsenicals have included the copper arsenate, Paris green, lead arsenate, and calcium arsenate. The arsenicals uncouple oxidative phosphorylation, inhibit certain enzymes that contain sulfhydryl (-SH) groups, and coagulate protein by causing the shape or configuration of proteins to change.

The inorganic fluorides were sodium fluoride, barium fluosilicate, sodium silicofluoride, and cryolite. Cryolite has returned in recent years as a relatively safe fruit and vegetable insecticide, used in integrated pest management Programmes. The fluoride ion inhibits many enzymes that contain iron, calcium, and magnesium. Several of these enzymes are involved in energy production in cells, as in the case of phosphatases and phosphorylases.

Boric acid, used against cockroaches and other crawling household pests in the 1930's and '40's, has also returned. As a salt, it is non-volatile and will remain effective as long as it is kept dry and in adequate concentration.

Consequently, it has the longest residual activity of any insecticide used for crawling household insects, and is quite useful in the control of all cockroach species when placed in wall voids and other protected, difficult-to-reach sites. It acts as a stomach poison and insect cuticle wax absorber.

Sodium borate (disodium octaborate tetrahydrate) resembles boric acid in its action. This water-soluble salt is used to treat lumber and other wood products to control decay fungi, termites, and other wood infesting pests. The last group of inorganics is the silica gels or silica aerogels– light, white, fluffy, silicate dusts used for household insect control.

The silica aerogels kill insects by absorbing waxes from the insect cuticle, permitting the continuous loss of water from the insect body, causing the insects to become desiccated and die from dehydration. These include Dri-Die, Drianone, and Silikil Microcel. Drianone is fortified with pyrethrum and synergists to enhance its effectiveness.

NEW-MISC. INSECTICIDE CLASSES

Seven classes of insecticides have made their appearance in recent years. These are summarized below.

METHOXYACRYLATES

Fluacrypyrin is an acaricide for fruit and is the only example of this class currently. It is registered for use on fruit in Japan.

Naphthoquinones

Acequinocyl is a miticide with insecticidal activity for pome fruit, nut crops, citrus and ornamentals. It is the only member of this group at present and the mode of action is not yet determined. It holds registrations in Korea and Japan but not in the US.

Acequinocyl (Kanemite, Piton):

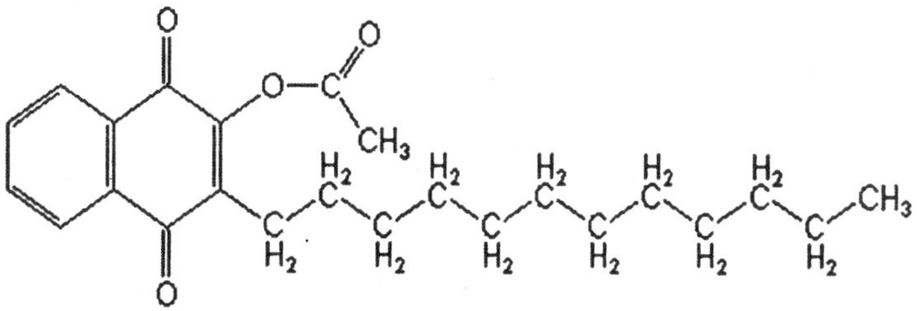

3-dodecyl-1,4-dihydro-1,4-dioxo-2-naphthyl acetate

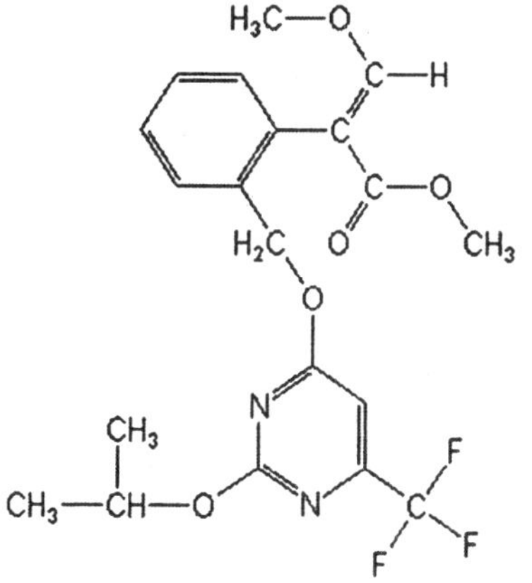

Methyl(*E*)-2-{a-[2-isopropoxy-6-(trifluoromethyl)pyrimidin-4-yloxy]-o-tolyl}-3-methoxyacrylate

NEREISTOXIN ANALOGUES

It includes thiocyclam, cartap, bensultap, and thiocytap-sodium. Analogues of nereistoxin have been known for decades. They generally are stomach poisons with some contact action and often show some systemic action. A major share of the development and use of these compounds has taken place in Japan. They are based on a natural toxin of the marine worm *Lumbriconereis heteropoda*. Of the many analogs synthesized only those that were metabolized back to the original nereistoxin after application were active.

In this sense members of this class are *proinsecticides* in that they are applied in their manufactured form but are known to degrade to a specific active component. The members of this group tend to be selectively active on Colopteran and Lepidopteran insect pests.

Cartap is a broad spectrum insecticide with good activity against rice stem borer. Bensultap is used to control the Colourado Potato beetle and other insect pests. Thiosultap-sodium is used to control selected beetle and Lepidopteran pests on rice, vegetables and fruit trees.

Thiocyclam, is used for the control of similar pests in several crops. Members of this class act as acetyl choline receptor agonists at low concentrations and as channel blockers at higher concentrations.

Although there has been commercial interest in thiocyclam for use in the US we do not believe there are commercial examples that are to achieve U.S. registration.

Thiocyclam (Evisect):

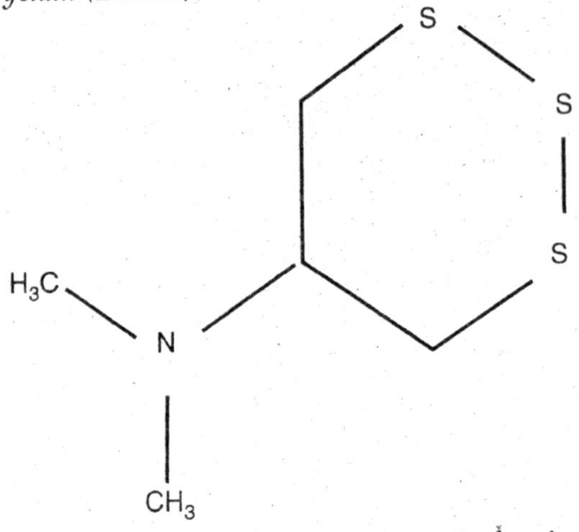

N,N-dimethyl-1,2,3-trithian-5-ylamine

Pyridine Azomethine

Pymetrozine, first registered in 1999 by EPA has a unique mode of action that is not fully understood. It appears to act by preventing insects from the Order Homoptera from inserting their stylus into plant tissue. Pymetrozine is used to control aphids and whiteflies in vegetables, potatoes, tobacco, deciduous citrus fruit hops and ornamentals.

Pymetrozine (Fulfill):

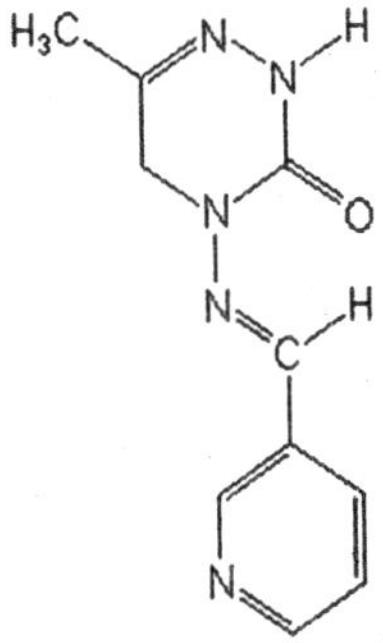

(*E*)-4,5-dihydro-6-methyl-4-(3-pyridylmethyleneamino)-1,2,4-triazin-3(2*H*)-one

PYRIMIDINAMINES

Pyrimidifen (Miteclean) is an insecticide and miticide. As a miticide the product controls spider and rust mites in deciduous fruits, citrus, vegetables and tea. As an insecticide it controls diamondback moth in vegetables. Very little information is available on the other member of this class, Flufenerim (S-1560), other than it is insecticidal.

Pyrimidifen *(Miteclean)*:

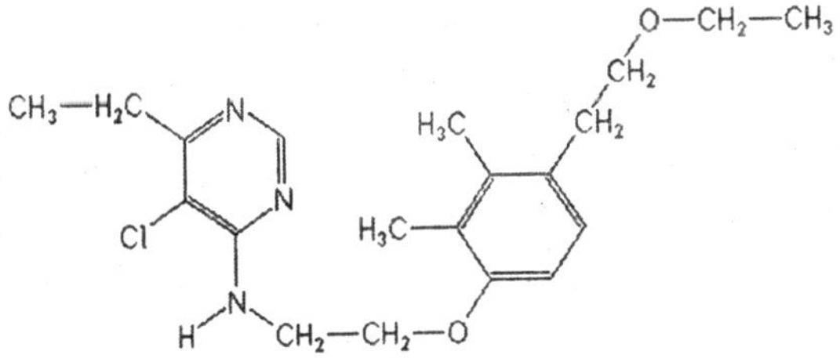

5-chloro-*N*-{2-[4-(2-ethoxyethyl)-2,3 dimethylphenoxy] -ethyl} -6-ethylpyrimidin-4-amine

TETRONIC ACIDS

Spirodiclofen and spiromesifen are the only two members of this recently introduced class. Spirodiclofen has broad-spectrum activity against mites, and controls scale crawlers and psyllad nymphs. Action is good on eggs and quiescent stages. Target crops are citrus, grapes, nuts, pome and stone fruits.

Spirodiclofen:

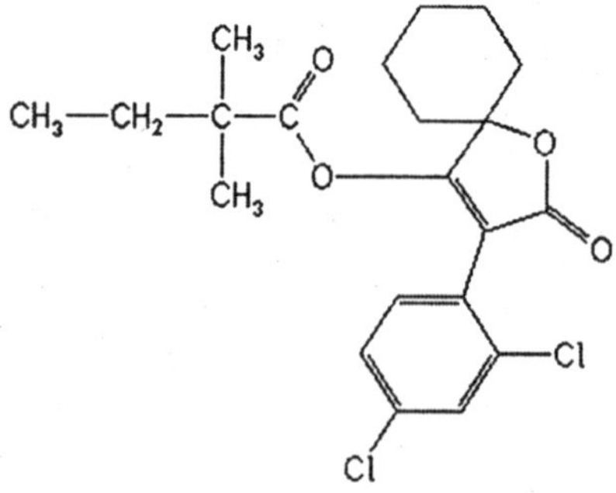

3-(2,4-dichlorophenyl)-2-oxo-1-oxaspiro dec-3-en-4-yl 2,2-dimethylbutyrate

MISCELLANEOUS COMPOUNDS

Clofentezine belongs to the unique group, the tetrazines, used as an acaricide/ovicide for deciduous fruits, citrus, cotton, cucurbits, vines and ornamentals. A newer and somewhat similar agent is etoxazole which is an acaricide registered for ornamentals grown in greenhouses. The modes of action of these two compounds are not yet understood.

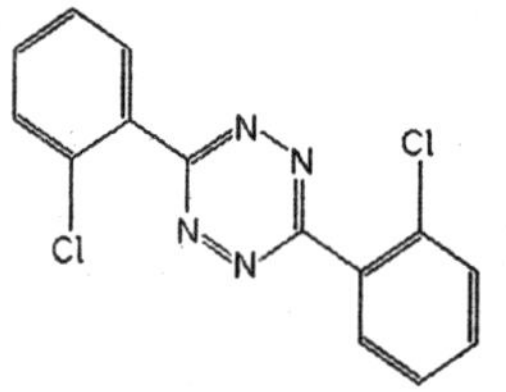

Etoxazole:

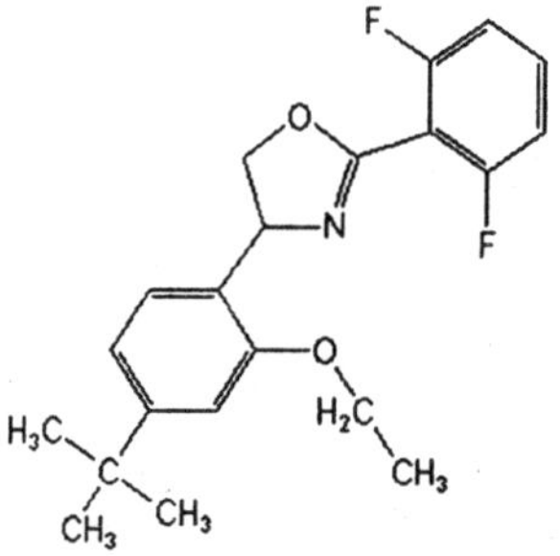

(*RS*)-5-tert-butyl-2-[2-(2,6-difluorophenyl)-4,5-dihydro-1,3-oxazol-4-yl]phenetole

Enzone, sodium tetrathiocarbonate, is used only on grapes and citrus applied as a water application and irrigated into the soil. It breaks down in the soil to form carbon disulfide, which acts rapidly, decomposes quickly, and is effective against nematodes, soil insects, and soil borne diseases.

Pyridalyl (S-1812):

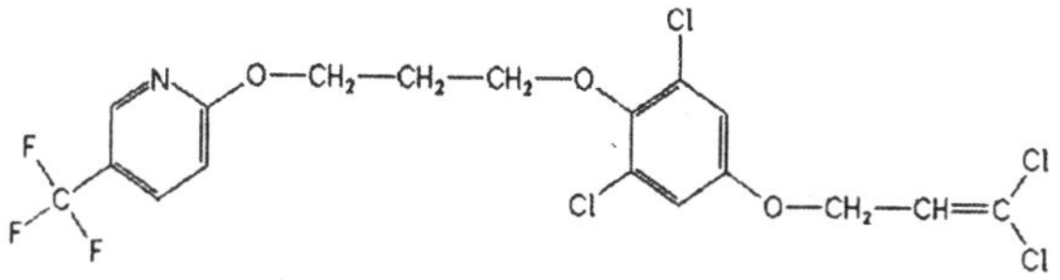

2,6-dichloro-4-(3,3-dichloroallyloxy)phenyl 3-[5-(trifluoromethyl)-2-pyridyloxy]propyl ether

Amidoflumet (S-1955):

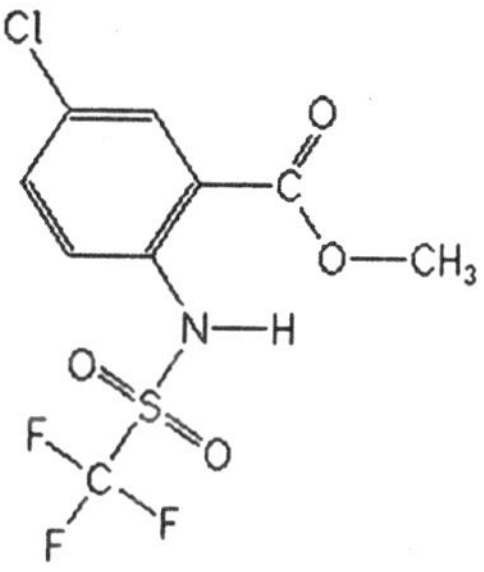

Methyl5-chloro-2-[(trifluoromethyl) sulfonyl] amino} benzoate. The newest agents in this category are pyridanyl and amidoflumet. Pryidalyl (S-1812) is active on Lepidoptera and thrips and has the advantage of being active against pyrethroid-resistant insects. Little more is available on amidoflumet (S-1955) other than it is an acaricide in the early stage of its development.

BIORATIONAL INSECTICIDES

The U.S. EPA identifies biorational pesticides as inherently different from conventional pesticides, having fundamentally different modes of action, and consequently, lower risks of adverse effects from their use. Biorational has come to mean any substance of natural origin (or man-made substances resembling those of natural origin), that has a detrimental or lethal effect on specific target pest(s), e.g., insects, weeds, plant diseases (including nematodes), and vertebrate pests, possess a unique mode of action, are non-toxic to man and his domestic plants and animals, and have little or no adverse effects on wildlife and the environment. EPA uses a similar term, biopesticides, which will be defined below.

Biorational insecticides are grouped as either:

- Biochemicals (hormones, enzymes, pheromones and natural agents, such as insect and plant growth regulators),
- Microbial (viruses, bacteria, fungi, protozoa, and nematodes).

In the 1990's the US-EPA began to emphasize a class of products known as biopesticides. EPA places biopesticides into three categories

- Microbial pesticides (bacteria, fungi, virus or protozoa)
- Biochemicals – natural substances that control pests by non-toxic mechanisms. An example is insect pheromones.
- Plant-Incorporated protectants (PIPs) – (primarily transgenic plants, e.g., Bt corn).

EPA discloses that at the end of 2001 there were nearly 200 biopesticide active ingredients registered comprising nearly 800 products.

Characteristics that distinguish biorational and biopesticides from conventional ones include very low orders of toxicity to non-target species, pest targets are specific, generally low use rates, rapid decomposition in the environment, usually work well in IPM Programmes and reduce reliance on conventional pesticide products.

The terms "biorational" and "biopesticide" overlap but are not identical. Below is an overview of what is considered as biorational insecticides. In some cases there are overlaps with botanicals (e.g., rotenone, florals, etc. and also conventional insecticides (e.g., benzoylureas). We will point out the discrepancies in classification between the biorational and biopesticide categories where they occur.

INSECT PHEROMONES

Most insects appear to communicate by releasing molecular quantities of highly specific compounds that vapourize readily and are detected by insects of the same species. These delicate molecules are known as pheromones. The word pheromone comes from the Greek pherein, "to carry," and hormon, "to excite or stimulate."

Of 1,314 species of insects with confirmed attraction responses to identified pheromones, 1,260 of these pheromones are produced by females. Only 54 species use male-produced sex attractants. In a few species both sexes produce the same attractant by both sexes.

Pheromones are classified as either, releasers and or primers. Releasers are fast-acting and are used by insects for sexual attraction, aggregation (including trail following), dispersion, oviposition, and alarm. Primers are slow-acting and cause gradual changes in growth and development, especially in social insects by regulating caste ratios of the colony.

The five principal uses for sex pheromones are:

- Male trapping, to reduce the reproductive potential of an insect population;
- Movement studies, to determine how far and where insects move from a given point;
- Population monitoring, to determine when peak emergence or appearance occurs;
- Detection Programmes, to determine if a pest occurs in a limited trapping area, such as around international airports or quarantined areas;
- The "confusion or mating disruption" technique.

The first use of mating disruption involved gossyplure, the pink bollworm pheromone. Incorporated into small, hollow, polyvinyl fibres that permit slow release of the pheromone, it was broadcast heavily and uniformly over infested cotton fields. In mid 2002, EPA had registered 36 pheromones which comprised over 200 individual products.

Despite praise for the potential of sex pheromones, they are most practically used in survey traps to provide information about population levels, to delineate infestations, to monitor control or eradication Programmes, and to warn of new pest introductions.

INSECT GROWTH REGULATORS

Insect growth regulators (IGRs) are chemical compounds that alter growth and development in insects. The IGRs disrupt insect growth and development in three ways As juvenile hormones, as precocenes, and as chitin synthesis inhibitors.

Juvenile hormones (JH) include ecdysone (the molting hormone), JH mimic, JH analog (JHA), and are known by their broader synonyms, juvenoids and juvegens. They disrupt immature development and emergence as adults. Precocenes interfere with the normal function of glands that produce juvenile hormones. And, chitin synthesis inhibitors, (conventional benzoylureas, buprofezin and cyromazine), affect the ability of insects to produce new exoskeletons when molting.

The IGRs are effective when applied in very minute quantities and generally have few or no effects on humans and wildlife. They are, however, nonspecific, since they affect not only the target species, other arthropods as well.

Instead of killing directly, IGRs interfere in the normal mechanisms of development and cause the insects to die before reaching the adult stage.

One JH is the classical juvabione, found in the wood of balsam fir. Its effect was discovered quite by accident when paper towels made from this source were used to line insect-rearing containers, and the insects' development was suppressed.

Some of these plant-derived substances actually serve to inhibit the development of insects feeding thereon, thus protecting the host plant. These are referred to broadly as antijuvenile hormones, more accurately, antiallatotropins, or precocenes. Although the mode of action of the precocenes is still unclear, it is known that they depress the level of juvenile hormone below that normally found in immature insects.

For practical purposes, IGRs are used on crops to suppress damaging insect numbers. They would be applied with the purpose of preventing pupal development or adult emergence, thus keeping the insects in the immature stages, resulting eventually in their deaths. Commercial successful pheromones have shown activity on mosquito larvae, caterpillars, and hemipterans (bugs), although effects have been observed on practically all insect orders.

Methoprene (Altosid):

$$CH_3-C(CH_3)(O-CH_3)-CH_2-CH_2-CH_2-CH(CH_3)-CH_2-CH=CH-C(CH_3)=CH-C(=O)-O-CH(CH_3)-CH_3$$

- Methylethyl (2E,4E)-11-methoxy-3,7,11-trimethyl-2,4-dodecadienoate

Fenoxycarb is a carbamate stomach insecticide that has also JH-type effects when contacted or ingested by a wide array of arthropod pests, e.g., ants, roaches, ticks, chiggers and many others.

Pyriproxifen is an effective molt inhibitor for a wide range of insects, but particularly useful for whitefly on cotton, citrus scales, fly-breeding sites such as livestock and poultry houses, and aquatic sites for mosquito control. Another is buprofezin classed as a thiadiazine IGR. Both have given excellent results in controlling the whitefly complex, now a universal problem in U. S. cotton production.

Pyriproxifen:

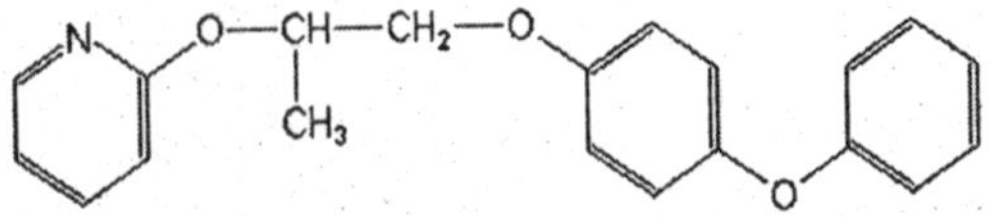

2-[1-methyl-2-(4-phenoxyphenoxy)ethoxy]pyridine

Buprofezin:

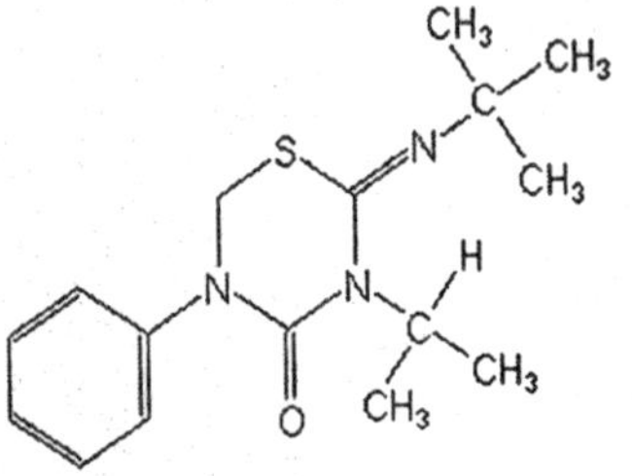

2-[(1,1-dimethylethyl)imino]tetrahydro-3-(1-methylethyl)-5-phenyl-4*H*-1,3,5-thiadiazin-4-one

None of the above– pyriproxifen, buprofezin, fenoxycarb or the methoprene, hydroprene group of juvenile hormone mimics are considered to be biopesticides by EPA.

HYDRAZINE INSECTICIDE/IGRS

A newer class of insecticidal IGRs is the hydrazines, which includes tebufenozide, halofenozide, methoxyfenozide and chromafenozide. All are ecdysone agonists or disruptors. EPA has not classified members of this group as biopesticides. Tebufenozide, in addition to being both a stomach and contact insecticide, has also JH-IGR characteristics. It disrupts the molting process by antagonizing ecdysone, the molting hormone. Lepidopteran pests are controlled while maintaining natural populations of beneficial insect predators and parasites. Halofenozide registered in1999, is a systemic IGR, effective on cutworms, sod webworms, armyworms and white grubs, and has some ovicidal activity.

It lacks the stomach or contact characteristics of tebufenozide. Methoxyfenozide, like tebufenozide, is both a stomach and contact insecticide with JH-IGR qualities.

It is systemic only through the roots. Pests controlled are lepidopterans such as codling moth, oriental fruit moth, European corn borer, and others. Crop candidates are cotton, corn, vegetables, pome fruit, and grapes. EPA considered methoxyfenozide as a reduced-risk candidate and first registered it in mid-2000. Chromafenozide (Matric) is a newer member of this group, not registered in the U.S., and is used to control various lepidopteran pests in vegetables and ornamentals.

Tebufenozide:

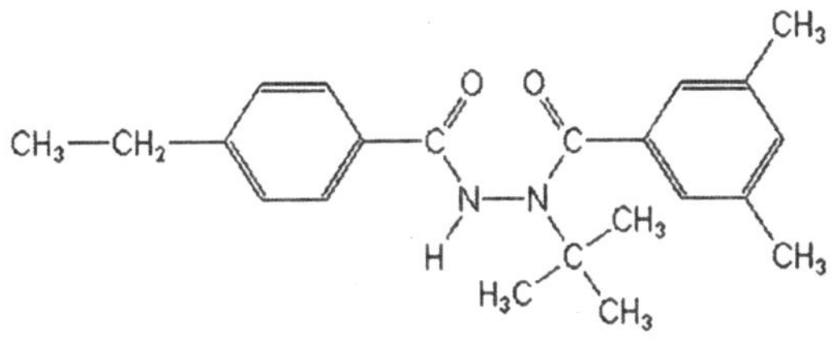

3,5-dimethylbenzoic acid 1-(1,1-dimethylethyl)-2-(4-ethylbenzoyl)hydrazide

Other Biorational Insecticides

A number of the products that we covered under botanicals and florals are also considered by many to be biorational products, and indeed, EPA includes them under the biopesticide category. Some examples include Neem oil, cinnamaldehyde, and eugenol. A new product, Virtuoso, is a Streptomycetes-based agent that controls caterpillars but little is yet published on it, at present.

Clandosan is a naturally occurring product derived from crab and shrimp shells and used as a nematicide. It is a dried, powdered, chitin protein isolated from crustacean exoskeletons and blended with urea. It stimulates growth of beneficial soil microorganisms that control nematodes, but does not have a direct adverse effect on nematodes as such.

MICROBIALS

Microbial insecticides obtain their name from

microorganisms that are used to control certain insects. The insect disease-causing microorganisms do not harm other animals or plants. At present there are relatively few produced commercially and approved by the EPA (over 55 natural, and 16 bioengineered organisms) for use on food and feed crops. In mid-2002, the EPA list of registered microbials included 35 bacteria, 1 yeast, 17 fungi, 1 protozoan, 6 viruses, 8 bioengineered organisms and 8 transgenic crop genes.

The insecticidal bacterium *Bacillus thuringiensis* (*Bt*) was discovered in the early 20th century. It occurs as a large number of subspecies that are identified among other characteristics by surface antigens, plasmid arrays, and breadth of species responding to its insecticidal action. *Bt* is a soil inhabiting, gram-positive sporulating bacterium that produces one or more very tiny parasporal crystals within its sporulating cells. These crystals are composed of large proteins known as delta-endotoxins. Delta-endotoxins act by binding to specific receptor sites on the gut epithelium, leading slowly to degradation of the gut lining and starvation. Thus, several days are required to kill insects that have ingested *Bt* products.

Over time, several *B. thuringiensis* varieties have been discovered, each with its distinct toxicity characteristics to different insect species. *B. thuringiensis* var. *kurstaki* was the first, being the spores and crystalline delta-endotoxin as the active ingredient, and produced by *B. thuringiensis* Berliner, var. *Kurstaki*, Serotype H-3a3b, HD-1, in fermentation. Products from this process control most lepidopteran pests, the caterpillars with high gut pH, which include the armyworms, cabbage looper, imported cabbage worm, gypsy moth, and spruce budworm. The next was *B. thuringiensis* var. *israelensis,* being the crystalline delta-endotoxin as the active ingredient, and produced by fermentation of *B. thuringiensis* Berliner, var. *israelensis,* Serotype H-14. These products are used primarily for the control of aquatic insects, the mosquitoes and black flies in their larval forms.

Then came *B. thuringiensis var. aizawai,* produced by this variety, Serotype H-7, in fermentation. This product is currently registered only for the control of the wax moth larval infestations in the honey comb of honey bees. Following this came *B. thuringiensis* var. *morrisoni,* spores and delta-endotoxin produced by fermentation of Serotype 8a8b. This is again a broad spectrum Bt for most caterpillars on most crops including the home garden. *B. thuringiensis* var. *san diego* was developed for Colourado potato beetle control on all its hosts, the elm leaf beetle and other beetle larvae on a wide range of shade and ornamental trees.

This was the first Bt product that was effective against coleopteran larvae. *B. thuringiensis* var. *tenebrionis* and the identical var. *san diego* were also developed for the Colourado potato beetle.The utilization of Bt genes transplanted into crops, which is addressed elsewhere in this document, is transforming the area of microbial pesticides

An innovative development in the agricultural use of microbial insecticides was the addition of feeding or gustatory stimulants, making the mixtures serves as baits. The feeding stimulants attracted the caterpillars to treated foliage, which increased their consumption of the microbial.

Fungi

Mycar was a promising biorational miticide, a mycoacaricide, but was discontinued by the manufacturer in 1984. The microorganism was *Hirsutella thompsonii,* a parasitic fungus that infects and kills the citrus rust mite.

Under optimum conditions *H. thompsonii* can infect spider mites and other nontarget mites. It was, however, consistently effective only against the citrus rust mite; thus a selective miticide.

EPA registered *Metarhizium anisopliae* St. F52 in mid-2002 to control various ticks, beetles, flies, gnats and thrips for non-food outdoor and greenhouse uses. Certain of the registered uses were conditional for two years pending results of tick performance studies.

Another strain of this organism is also registered as a termiticide. Application was made in 1998 to register the fungus *Aspergillus flavus* strain AF36 as a bioinsecticide for cotton. Its purpose is to help reduce the incidence of other *Aspergillus* spp. that produce the highly toxic mycotoxin, aflatoxin, in cotton seed.

PROTOZOA

Nosema locustae is a biorational originally developed by Sandoz, Inc., in 1981, for the control of grasshoppers. Marketed under the names, and Grasshopper Attack, the microorganism is a protozoan. These have been discontinued although the registrations remain.

NEMATODES

There were two commercial nematode products available for termite control. The nematode, *Neoaplectana carpocapsae,* in the family Steirnernamatidae, is specific for subterranean termites. It kills all stages of these termites by delivering a pathogenic bacterium, *Xenorhabdus* spp., which is lethal within 48 hours after penetration. Unfortunately, neither product succeeded commercially.

TRANSGENIC PESTICIDES

The concept that organisms and crops could be engineered to augment pest control was well known in the 1970's. By the mid 1980s, one company, Monsanto had committed to a research Programme designed to create crop protection products through the application of biotechnology. Charles has produced a very readable history of pesticide-related transgenic crops and this book is recommended to those who want to understand how this new technology unfolded.

Transgenic organisms are genetically altered by artificial introduction of DNA from another organism. The artificial gene sequence is referred to as a *transgene*. Plants with such transgenes are also referred to as being *genetically modified* (GM). Plants that emulate insecticides are those altered to induce *insect-resistance* (also called *plant pesticides* or *plant incorporated protectants*. The purpose of the following paragraphs is to summarize what biotechnology has contributed to insecticide science in the course of just the last decade or so.

Research built on the elucidation of the genetic code in the early 1950's and culminating in the 1990's allowed those using the techniques of biotechnology to move genes coding for specific traits from selected organism to crop cells. Such altered cells were then regenerated to viable crop plants through tissue culture. Several transgenic crops have thus been and are being created from backcrossing the selected traits into elite seed lines. The result has led us to *plant pesticides*.

Plant pesticides are defined by EPA as plants that have been genetically engineered to contain the delta-endotoxin genes from *Bacillus thuringiensis*. This definition will expand as genes from additional sources are incorporated into plants.

In 1995, EPA registered the first plant pesticide. It was Monsanto's Bt-cotton containing B. *t. Cry1Ac* delta-endotoxin, following more than a decade of research. This novel form of cotton was introduced experimentally in 1995 as Bollgard cotton, resistant to tobacco budworm, cotton bollworm, and pink bollworm with activity on other minor lepidopteran pests.

Bt-enhanced cotton, corn and other insect resistant crops produce one or more crystalline proteins that disrupt the gut lining of susceptible insect pests feeding on their tissues which cause the pests to stop feeding and die. Several plant pesticides have been introduced in the U.S. since 1995. Some of these have been very successful commercially while others, such as, NewLeaf Potatoes, have been withdrawn from the market.

In some subsequent product introductions the performance of these plant pesticides have been enhanced or augmented by use of *stacked genes*. This means that more than one transgene is introduced into the same crop to achieve multiple desired characteristics.

In the U.S. the proportion of total planted acres of *B.t.* corn, grew from 18% in 2000 to 26% in 2003, and this does not include stacked-gene plant insecticides. The proportions for cotton were even higher if stacked-gene varieties were included.

Three federal agencies regulate the release and use of transgenic plants and plant pesticides under a coordinated framework. There is a sharing and partitioning of biotechnology regulatory responsibility among these three agencies. The approach of each agency is similar, in that, they each evaluate risks from a sound science perspective and regulate individual products on a case-by-case basis. There is considerable reliance on comparing transgenic organisms with their conventional counterparts that have a known history of safe use. The emphasis during the regulatory review is to assure that the GM organism will not produce harmful toxins, allergens, etc., and will not, once released, cause adverse effects.

PESTICIDES

A pesticide is a substance or mixture of substances used for preventing, controlling, or lessening the damage caused by a pest. A pesticide may be a chemical substance, biological agent (such as a virus or bacteria), antimicrobial, disinfectant or device used against any pest. Pests include insects, plant pathogens, weeds, molluscs, birds, mammals, fish, nematodes (roundworms) and microbes that compete with humans for food, destroy property, spread or are a vector for disease or cause a nuisance.

Although there are benefits to the use of pesticides, there are also drawbacks, such as potential toxicity to humans and other animals.Pesticides are hazadous to some wildlife in the sea because it gets evapourated and goes into the clouds.Then it rains, surface run-off into the sea and poisions them.

Types of pesticides

There are multiple ways of classifying pesticides:

- Algicides or Algaecides for the control of algae

- Avicides for the control of birds
- Bactericides for the control of bacteria
- Fungicides for the control of fungi and oomycetes
- Herbicides for the control of weeds
- Insecticides for the control of insects - these can be Ovicides (substances that kill eggs), Larvicides (substances that kill larvae) or Adulticides (substances that kill adult insects)
- Miticides or Acaricides for the control of mites
- Molluscicides for the control of slugs and snails
- Nematicides for the control of nematodes
- Rodenticides for the control of rodents
- Virucides for the control of viruses (e.g. H5N1)

Pesticides can also be classed as synthetic pesticides or biological pesticides (biopesticides), although the distinction can sometimes blur. Broad-spectrum pesticides are those that kill an array of species, while narrow-spectrum, or selective pesticides only kill a small group of species.

A systemic pesticide moves inside a plant following absorption by the plant. With insecticides and most fungicides, this movement is usually upward (through the xylem) and outward. Increased efficiency may be a result. Systemic insecticides which poison pollen and nectar in the flowers may kill needed pollinators such as bees. Most pesticides work by poisoning pests.

USES, BENEFITS AND DRAWBACKS

Pesticides are used to control organisms which are considered harmful.For example, they are used to kill mosquitoes that can transmit potentially deadly diseases like west nile virus and malaria. They can also kill bees, wasps or ants that can cause allergic reactions. Insecticides can protect animals from illnesses that can be caused by parasites such as fleas.Pesticides can prevent sickness in humans that could be caused by mouldy food or diseased produce. Herbicides can be used to clear roadside weeds, trees and brush.

They can also kill invasive weeds in parks and wilderness areas which may cause environmental damage. Herbicides are commonly applied in ponds and lakes to control algae and plants such as water grasses that can interfere with activities like swimmng and fishing and cause the water to look or smell unpleasant. Uncontrolled pests such as termites and mould can damage structures such as houses.

Pesticides are used in grocery stores and food storage facilities to manage rodents and insects that infest food such as grain. Each use of a pesticide carries some associated risk. Proper pesticide use decreases these associated risks to a level deemed acceptable by pesticide regulatory agencies such as the EPA and PMRA. Pesticides can save farmers money by preventing crop losses to insects and other pests; in the US, farmers get an estimated four-fold return on money they spend on pesticides. One study found that not using pesticides reduced crop yields by about 10%.

DDT, sprayed on the walls of houses, has been used to fight malaria since the 1950s. Recent policy statements by the World Health Organization have given stronger support to this approach. Dr. Arata Kochi, WHO's malaria chief, said, "One of the best tools we have against malaria is indoor residual house spraying. Of the dozen insecticides WHO has approved as safe for house spraying, the most effective is DDT."However, since then, an October 2007 study has linked breast cancer from exposure to DDT prior to puberty.

Scientists estimate that DDT and other chemicals in the organophosphate class of pesticides have saved 7 million human lives since 1945 by preventing the transmission of diseases such as malaria, bubonic plague, sleeping sickness, and typhus. However, DDT use is not always effective, as resistance to DDT was identified in Africa as early as 1955, and by 1972 nineteen species of mosquito worldwide were resistant to DDT. A study for the World Health Organization in 2000 from Vietnam established that non-DDT malaria controls were significantly more effective than DDT use.

In the US, about a quarter of pesticides used are used in houses, yards, parks, golf courses, and swimming pools. About 70% of the pesticides sold in the US are used in agriculture. Since before 2500 BC, humans have utilized pesticides to protect their crops. The first known pesticide was elemental sulfur dusting used in Sumeria about 4,500 years ago. By the 15th century, toxic chemicals such as arsenic, mercury and lead were being applied to crops to kill pests. In the 17th century, nicotine sulfate was extracted from tobacco leaves for use as an insecticide. The 19th century saw the introduction of two more natural pesticides, pyrethrum which is derived from chrysanthemums, and rotenone which is derived from the roots of tropical vegetables.

In 1939, Paul Müller discovered that DDT was a very effective insecticide. It quickly became the most widely-used pesticide in the world.

In the 1940s manufacturers began to produce large amounts of synthetic pesticides and their use became widespread. Some sources consider the 1940s and 1950s to have been the start of the "pesticide era." Pesticide use has increased 50-fold since 1950 and 2.5 million tons (2.3 million metric tons) of industrial pesticides are now used each year. Seventy-five per cent of all pesticides in the world are used in developed countries, but use in developing countries is increasing.

In the 1960s, it was discovered that DDT was preventing many fish-eating birds from reproducing, which was a serious threat to biodiversity. Rachel Carson wrote the best-selling book *Silent Spring* about biological magnification. DDT is now banned in at least 86 countries, but it is still used in some developing nations to prevent malaria and other tropical diseases by killing mosquitoes and other disease-carrying insects.

REGULATION

In most countries, in order to sell or use a pesticide, it must be approved by a government agency. For example, in the United States, the Environmental Protection Agency (EPA) does so. Complex and costly studies must be conducted to indicate whether the material is safe to use and effective against the intended pest. During the registration process, a label is created which contains directions for the proper use of the material. Based on acute toxicity, pesticides are assigned to a Toxicity Class.

Some pesticides are considered too hazardous for sale to the general public and are designated restricted use pesticides. Only certified applicators, who have passed an exam, may purchase or supervise the application of restricted use pesticides. Records of sales and use are required to be maintained and may be audited by government agencies charged with the enforcement of pesticide regulations.

In Canada, over 140 municipalities and the entire province of Quebec have now placed restrictions on the cosmetic use of synthetic lawn pesticides as a result of health and environmental concerns. The Ontario provincial government promised on September 24, 2007 to also implement a province-wide ban on the cosmetic use of lawn pesticides, for protecting the public. Medical and environmental groups support such a ban. On April 22, 2008, the Provincial Government of Ontario announced that it will pass legislation that will prohibit, province-wide, the cosmetic use and sale of lawn and garden pesticides.

The Ontario legislation would also echo Massachusetts law requiring pesticide manufacturers to reduce the toxins they use in production. The Province of Prince Edward Island is also considering such legislation. On April 3, 2008, the Canadian Cancer Society released opinion poll results conducted by Ipsos Reid, which established that a clear majority of residents in the provinces of British Columbia and Saskatchewan want province-wide cosmetic lawn pesticide bans, and that the majority of respondents believe that cosmetic pesticides are a threat to their health.

Though pesticide regulations differ from country to country, pesticides and products on which they were used are traded across international borders. To deal with inconsistencies in regulations among countries, delegates to a conference of the United Nations Food and Agriculture Organization adopted an International Code of Conduct on the Distribution and Use of Pesticides in 1985 to create voluntary standards of pesticide regulation for different countries. The Code was updated in 1998 and 2002. The FAO claims that the code has raised awareness about pesticide hazards and decreased the number of countries without restrictions on pesticide use.

Two other efforts to improve regulation of international pesticide trade are the United Nations London Guidelines for the Exchange of Information on Chemicals in International Trade and the United Nations Codex Alimentarius Commission.

The former seeks to implement procedures for ensuring that prior informed consent exists between countries buying and selling pesticides, while the latter seeks to create uniform standards for maximum levels of pesticide residues among participating countries. Both initiatives operate on a voluntary basis.

Reading and following label directions is required by law in countries such as the US and in limited parts of the rest of the world.

In the US, the Federal Insecticide, Fungicide, and Rodenticide Act (FIFRA) was first passed in 1947, giving the United States Department of Agriculture responsibility for regulating pesticides. In 1972, FIFRA underwent a major revision and transferred responsibility of pesticide regulation to the Environmental Protection Agency and shifted emphasis to protection of the environment and public health.

One study found pesticide self-poisoning the method of choice in one third of suicides worldwide, and recommended, among other things, more restrictions on the types of pesticides that are most harmful to humans.

Environmental Effects

Pesticide use raises a number of environmental concerns. Over 98% of sprayed insecticides and 95% of herbicides reach a destination other than their target species, including non-target species, air, water, bottom sediments, and food. Pesticide drift occurs when pesticides suspended in the air as particles are carried by wind to other areas, potentially contaminating them. Pesticides are one of the causes of water pollution, and some pesticides are persistent organic pollutants and contribute to soil contamination.

Health Effects

Pesticides can present danger to consumers, bystanders, or workers during manufacture, transport, or during and after use. The American Medical Association recommends limiting exposure to pesticides and using safer alternatives

Particular uncertainty exists regarding the long-term effects of low-dose pesticide exposures. Current surveillance systems are inadequate to characterize potential exposure problems related either to pesticide usage or pesticide-related illnesses...Considering these data gaps, it is prudent...to limit pesticide exposures...and to use the least toxic chemical pesticide or non-chemical alternative.

Farmers and Workers

There have been many studies of farmers with the goal of determining the health effects of pesticide exposure. The World Health Organisation and the UN Environment Programme estimate that each year, 3 million workers in agriculture in the developing world experience severe poisoning from pesticides, about 18,000 of whom die. According to one study, as many as 25 million workers in developing countries may suffer mild pesticide poisoning yearly.

Organophosphate pesticides have increased in use, because they are less damaging to the environment and they are less persistent than organochlorine pesticides. These are associated with acute health problems for workers that handle the chemicals, such as abdominal pain, dizziness, headaches, nausea, vomiting, as well as skin and eye problems. Additionally, many studies have indicated that pesticide exposure is associated with long-term health problems such as respiratory problems, memory disorders, dermatologic conditions, cancer, depression, neurological deficits, miscarriages, and birth defects. Summaries of peer-reviewed research have examined the link between pesticide exposure and neurologic outcomes and cancer, perhaps the two most significant things resulting in organophosphate-exposed workers.

There are concerns that pesticides used to control pests on food crops are dangerous to people who consume those foods. These concerns are one reason for the organic food movement. Many food crops, including fruits and vegetables, contain pesticide residues after being washed or peeled. Chemicals that are no longer used but which are resistant to breakdown for long periods may remain in soil and water and thus in food. The United Nations Codex Alimentarius Commission has recommended international standards for Maximum Residue Limits (MRLs), for individual pesticides in food. In the EU, MRLs are set by DG-SANCO. In the US, levels of residues that remain on foods are limited to tolerance levels that are established by the US EPA and are considered safe. The EPA sets the tolerances based on the toxicity of the pesticide and its breakdown products, the amount and frequency of pesticide application, and how much of the pesticide (i.e., the residue) remains in or on food by the time it is marketed and prepared.

Tolerance levels are obtained using scientific risk assessments that pesticide manufacturers are required to produce by conducting toxicological studies, exposure modeling and residue studies before a particular pesticide can be registered, however, the effects are tested for single pesticides, and there is little information on possible synergistic effects of exposure to multiple pesticide traces in the air, food and water.

A study published by the United States National Research Council in 1993 determined that for infants and children, the major source of exposure to pesticides is through diet. A study in 2006 measured the levels of organophosphorus pesticide exposure in 23 school children before and after replacing their diet with organic food (food grown without synthetic pesticides). In this study it was found that levels of organophosphorus pesticide exposure dropped dramatically and immediately when the children switched to an organic diet.

In the US, the National Academy of Sciences estimates that between 4,000 and 20,000 cases of cancer are caused per year by pesticide residues in food in allowable amounts.

The Pesticide Data Programme, a Programme started by the United States Department of Agriculture is the largest tester of pesticide residues on food sold in the United States. It began in 1991, and has since tested over 60 different types of food for over 400 different types of pesticides - with samples collected close to the point of consumption. Their most recent summary results are from the year 2005

To reduce the amounts of pesticide residues in food, consumers can wash, peel, and cook their food; trim the fat from meat; and eat a variety of foods to avoid repeat exposure to a pesticide typically us d on a given crop. Consumers can also buy food that is grown organically, though even organic food may have traces of pesticides. Exposure routes other than consuming food that contains residues, in particular pesticide drift, are potentially significant to the general public.

The Bhopal disaster occurred when a pesticide plant released 40 tons of methyl isocyanate (MIC) gas, intermediate chemical in the production of some pesticides. The disaster immediately killed nearly 3,000 people and ultimately caused at least 15,000 deaths. In China, an estimated half million people are poisoned by pesticides each year, 500 of whom die.

Children have been found to be especially susceptible to the harmful effects of pesticides. A number of research studies have found higher instances of brain cancer, leukemia and birth defects in children with early exposure to pesticides, according to the Natural Resources Defence Council.

Peer-reviewed studies now suggest neurotoxic effects on developing animals from organophosphate pesticides at legally-tolerable levels, including fewer nerve cells, lower birth weights, and lower cognitive scores.The EPA finished a 10 year review of the organophosphate pesticides following the 1996 Food Quality Protection Act, but did little to account for developmental neurotoxic effects, drawing strong criticism from within the agency and from outside researchers.

Some scientists think that exposure to pesticides in the uterus may have negative effects on a fetus that may manifest as problems such as growth and behavioural disorders or reduced resistance to pesticide toxicity later in life.

A new study conducted by the Harvard School of Public Health in Boston, has discovered a 70% increase in the risk of developing Parkinson's disease for people exposed to even low levels of pesticides. A 2008 study from Duke University found that the Parkinson's patients were 61 per cent more likely to report direct pesticide application than were healthy relatives.

Both insecticides and herbicides significantly increased the risk of Parkinson's disease.One study found that use of pesticides may be behind the finding that the rate of birth defects such as missing or very small eyes is twice as high in rural areas as in urban areas. Another study found no connection between eye abnormalities and pesticides. Pyrethrins, insecticides commonly used in common bug killers, can cause a potentially deadly condition if breathed in. Pesticide safety education and pesticide applicator regulation are designed to protect the public from pesticide misuse, but do not eliminate all misuse. Reducing the use of pesticides and choosing less toxic pesticides may reduce risks placed on society and the environment from pesticide use. Integrated pest management, the use of multiple approaches to control pests, is becoming widespread and has been used with success in countries such as Indonesia, China, Bangladesh, the US, Australia, and Mexico.

IPM attempts to recognize the more widespread impacts of an action on an ecosystem, so that natural balances are not upset. New pesticides are being developed, including biological and botanical derivatives and alternatives that are thought to reduce health and environmental risks. In addition, applicators are being encouraged to consider alternative controls and adopt methods that reduce the use of chemical pesticides. Pesticides can be created that are targeted to a specific pest's life cycle, which can be more environmentally-friendly. For example, potato cyst nematodes emerge from their protective cysts in response to a chemical excreted by potatoes; they feed on the potatoes and damage the crop. A similar chemical can be applied to fields early, before the potatoes are planted, causing the nematodes to emerge early and starve in the absence of potatoes.

Alternatives

Alternatives to pesticides are available and include methods of cultivation, use of other organisms to kill pests, genetic engineering, and methods of interfering with insect breeding. Cultivation practices include polyculture (growing multiple types of plants), crop rotation, planting crops in areas where the pests that damage them do not live, timing planting according to when pests will be least problematic, and use of trap crops that attract pests away from the real crop.

In the US, farmers have had success controlling insects by spraying with hot water at a cost that is about the same as pesticide spraying. Release of other organisms that fight the pest is another example of an alternative to pesticide use. These organisms can include natural predators or parasites of the pests. Biological pesticides based on entomopathogenic fungi, bacteria and viruses cause disease in the pest species can also be used. Interfering with insects' reproduction can be accomplished by sterilizing males of the target species and releasing them, so that they mate with females but do not produce offspring. This technique was first used on the screwworm fly in 1958 and has since been used with the medfly, the tsetse fly, and the gypsy moth. However, this can be a costly, time consuming approach that only works on some types of insects. Some evidence shows that alternatives to pesticides can be equally effective as the use of chemicals. For example, Sweden has halved its use of pesticides with hardly any reduction in crops. In Indonesia, farmers have reduced pesticide use on rice fields by 65% and experienced a 15% crop increase.

HERBICIDE

A herbicide is used to kill unwanted plants. Selective

herbicides kill specific targets while leaving the desired crop relatively unharmed. Some of these act by interfering with the growth of the weed and are often based on plant hormones.

Herbicides used to clear waste ground are nonselective and kill all plant material with which they come into contact. Some plants produce natural herbicides, such as the genus Juglans (walnuts). They are applied in total vegetation control (TVC) Programmes for maintenance of highways and railroads. Smaller quantities are used in forestry, pasture systems, and management of areas set aside as wildlife habitat.

Herbicides are widely used in agriculture and in landscape turf management. In the U.S., they account for about 70% of all agricultural pesticide use. Prior to the widespread use of chemical herbicides, cultural controls, such as altering soil pH, salinity, or fertility levels, were used to control weeds. Mechanical control (including tillage) was also (and still is) used to control weeds.

The first widely used herbicide was 2,4-dichlorophenoxyacetic acid, often abbreviated 2,4-D. It was first commercialized by the Sherwin-Williams Paint company and saw use in the late 1940s. It is easy and inexpensive to manufacture, and kills many broadleaf plants while leaving grasses largely unaffected (although high doses of 2,4-D at crucial growth periods can harm grass crops such as maize or cereals).

The low cost of 2,4-D has led to continued usage today and it remains one of the most commonly used herbicides in the world. Like other acid herbicides, current formulations utilize either an amine salt (usually trimethylamine) or one of many esters of the parent compound. These are easier to handle than the acid.

2,4-D exhibits relatively good *selectivity*, meaning, in this case, that it controls a wide number of broadleaf weeds while causing little to no injury to grass crops at normal use rates. A herbicide is termed selective if it affects only certain types of plants, and nonselective if it inhibits a very broad range of plant types. Other herbicides have been more recently developed that achieve higher levels of selectivity than 2,4-D.

The 1950s saw the introduction of the triazine family of herbicides, which includes atrazine, which have current distinction of being the herbicide family of greatest concern regarding groundwater contamination. Atrazine does not break down readily (within a few weeks) after being applied to soils of above neutral pH.

Under alkaline soil conditions atrazine may be carried into the soil profile as far as the water table by soil water following rainfall causing the aforementioned contamination. Atrazine is said to have *carryover,* a generally undesirable property for herbicides.

Glyphosate, frequently sold under the brand name Roundup, was introduced in 1974 for non-selective weed control. It is now a major herbicide in selective weed control in growing crop plants due to the development of crop plants that are resistant to it. The pairing of the herbicide with the resistant seed contributed to the consolidation of the seed and chemistry industry in the late 1990s.

Many modern chemical herbicides for agriculture are specifically formulated to decompose within a short period after application. This is desirable as it allows crops which may be affected by the herbicide to be grown on the land in future seasons. However, herbicides with low residual activity (i.e., that decompose quickly) often do not provide season-long weed control.

HEALTH EFFECTS

Certain herbicides affect metabolic pathways and systems unique to plants and not found in animals making many modern herbicides among the safest crop protection products having essentially no effect on mammals, birds, amphibians or reptiles.

Some herbicides can cause a variety of health effects ranging from skin rashes to death. The pathway of attack can arise from intentional or unintentional direct consumption of the herbicide, improper application resulting in the herbicide coming into direct contact with people or wildlife, inhalation of aerial sprays, or food consumption prior to the labeled pre-harvest interval. Under extreme conditions herbicides can also be transported via surface runoff to contaminate distant water sources. Most herbicides decompose rapidly in soils via soil microbial decomposition, hydrolysis or photolysis and some herbicides are more persistent with longer soil half-lives.

Other alleged health effects can include chest pain, headaches, nausea and fatigue. All organic and non-organic herbicides must be extensively tested prior to approval for commercial sale and labeling by the Environmental Protection Agency. However, because of the large number of herbicides in use, there is significant concern regarding health effects. Some of the herbicides in use are known to be mutagenic, carcinogenic or teratogenic.

However, some herbicides may also have a therapeutic use. Current research aims to use herbicides as an anti-malaria drug that targets the plant-like apicoplast plastid in the malaria-causing parasite *Plasmodium falciparum*.

CLASSIFICATION OF HERBICIDES

Herbicides can be grouped by activity, use, chemical family, mode of action, or type of vegetation controlled.

By activity:

- Contact herbicides destroy only the plant tissue in contact with the chemical. Generally, these are the fastest acting herbicides. They are less effective on perennial plants, which are able to regrow from rhizhomes, roots or tubers.
- Systemic herbicides are translocated through the plant, either from foliar application down to the roots, or from soil application up to the leaves. They are capable of controlling perennial plants and may be slower acting but ultimately more effective than contact herbicides.

By use:

- Soil-applied herbicides are applied to the soil and are taken up by the roots of the target plant.
- Pre-plant incorporated herbicides are soil applied prior to planting and mechanically incorporated into the soil.
- Preemergent herbicides are applied to the soil before the crop emerges and prevent germination or early growth of weed seeds.
- Post-emergent herbicides are applied after the crop has emerged.

Their classification by mechanism of action (MOA) indicates the first enzyme, protein, or biochemical step affected in the plant following application.

The main mechanisms of action are:

- ACCase inhibitors are compounds that kill grasses. Acetyl coenzyme A carboxylase (ACCase) is part of the first step of lipid synthesis. Thus, ACCase inhibitors affect cell membrane production in the meristems of the grass plant. The ACCases of grasses are sensitive to these herbicides, whereas the ACCases of dicot plants are not.
- ALS inhibitors the acetolactate synthase (ALS) enzyme (also known as acetohydroxyacid synthase, or AHAS) is the first step in the synthesis of the branched-chain amino acids (valine, leucine, and isoleucine). These herbicides slowly starve affected plants of these amino acids which eventually leads to inhibition of DNA synthesis. They affect grasses and dicots alike. The ALS inhibitor family includes sulfonylureas (SUs), imidazolinones (IMIs), triazolopyrimidines (TPs), pyrimidinyl oxybenzoates (POBs), and sulfonylamino carbonyl triazolinones (SCTs). ALS is a biological pathway that exists only in plants and not in animals thus making the ALS-inhibitors among the safest herbicides.
- *EPSPS inhibitors*: The enolpyruvylshikimate 3-phosphate synthase enzyme EPSPS is used in the

synthesis of the amino acids tryptophan, phenylalanine and tyrosine. They affect grasses and dicots alike. Glyphosate (Roundup) is a systemic EPSPS inhibitor but inactivated by soil contact.

- Synthetic auxin inaugurated the era of organic herbicides. They were discovered in the 1940s after a long study of the plant growth regulator auxin. Synthetic auxins mimic this plant hormone. They have several points of action on the cell membrane, and are effective in the control of dicot plants. 2,4-D is a synthetic auxin herbicide.
- Photosystem II inhibitors reduce electron flow from water to NADPH2+ at the photochemical step in photosynthesis. They bind to the Qb site on the D1 protein, and prevent quinone from binding to this site. Therefore, this group of compounds cause electrons to accumulate on chlorophyll molecules. As a consequence, oxidation reactions in excess of those normally tolerated by the cell occur, and the plant dies. The triazine herbicides (including atrazine) and urea derivatives (diuron) are photosystem II inhibitors.

ORGANIC HERBICIDES

Almost all herbicides in use today are considered "organic" herbicides in that they contain carbon as a primary molecular component.

An notable exception would be the arsenical class of herbicides. Sometimes they are referred to as synthetic organic herbicides. Recently the term "organic" has come to imply products used in organic farming.

Under this definition an organic herbicide is one that can be used in a farming enterprise that has been classified as organic.

Organic herbicides are expensive and may not be affordable for commercial production. They are much less effective than synthetic herbicides and are generally used along with cultural and mechanical weed control practices.

Organic herbicides include:

- Spices are now effectively used in patented herbicides.
- Vinegar is effective for 5-20% solutions of acetic acid with higher concentrations most effective but mainly destroys surface growth and so respraying to treat regrowth is needed. Resistant plants generally succumb when weakened by respraying.
- Steam has been applied commercially but is now considered uneconomic and inadequate. It kills surface growth but not underground growth and so respraying to treat regrowth of perennials is needed.
- Flame is considered more effective than steam but suffers from the same difficulties.

APPLICATION

Most herbicides are applied as water-based sprays using ground equipment. Ground equipment varies in design, but large areas can be sprayed using self-propelled sprayers equipped with a long boom, of 60 to 80 feet (20 to 25 m) with flat fan nozzles spaced about every 20 in (500 mm). Towed, handheld, and even horse-drawn sprayers are also used.

Synthetic organic herbicides can generally be applied aerially using helicopters or airplanes, and can be applied through irrigation systems (chemigation).

Terminology

- Control is the destruction of unwanted weeds, or the damage of them to the point where they are no longer competitive with the crop.
- Suppression is incomplete control still providing some economic benefit, such as reduced competition with the crop.
- Crop Safety, for selective herbicides, is the relative absence of damage or stress to the crop. Most selective herbicides cause some visible stress to crop plants.

Major Herbicides in use Today

- 2,4-D, a broadleaf herbicide in the phenoxy group used in turf and in no-till field crop production. Now mainly used in a blend with other herbicides that allow lower rates of herbicides to be used, it is the most widely used herbicide in the world, third most commonly used in the United States. It is an example of synthetic auxin(plant hormone).
- Atrazine, a triazine herbicide used in corn and sorghum for control of broadleaf weeds and grasses. Still used because of its low cost and because it works extrodinarily well on a broad spectrum of weeds common in the U.S. corn belt, Atrazine is commonly used with other herbicides to reduce the over-all rate of atrazine and to lower the for potential groundwater contamination, it is a photosystem II inhibitor.
- Clopyralid is a broadleaf herbicide in the pyridine group, used mainly in turf, rangeland, and for control of noxious thistles. Notorious for its ability to persist in compost. It is another example of synthetic auxin.
- Dicamba, a persistent broadleaf herbicide active in the soil, used on turf and field corn. It is another example of synthetic auxin.
- Glufosinate ammonium, a broad-spectrum contact herbicide and is used to control weeds after the crop emerges or for total vegetation control on land not used for cultivation.
- Glyphosate, a systemic nonselective (it kills any type of plant) herbicide used in no-till burndown and for weed control in crops that are genetically modified to resist its effects. It is an example of an EPSPs inhibitor.
- Imazapyr, is a non-selective herbicide used for the control of a broad range of weeds including

terrestrial annual and perennial grasses and broadleaved herbs, woody species, and riparian and emergent aquatic species.

- Imazapic, is a selective herbicide for both the pre- and post-emergent control of some annual and perennial grasses and some broadleaf weeds. Imazapic kills plants by inhibiting the production of branched chain amino acids (valine, leucine, and isoleucine), which are necessary for protein synthesis and cell growth.
- Linuron, is a non-selective herbicide used in the control of grasses and broadleafed weeds. It works by inhibiting photosynthesis.
- Metoalachlor, a pre-emergent herbicide widely used for control of annual grasses in corn and sorghum; it has largely replaced atrazine for these uses.
- Paraquat, a nonselective contact herbicide used for no-till burndown and in aerial destruction of marijuana and coca plantings. More acutely toxic to people than any other herbicide in widespread commercial use.
- picloram, a pyridine herbicide mainly used to control unwanted trees in pastures and edges of fields. It is another synthetic auxin.
- Triclopyr, a systemic, foliar herbicide in the pyridine group. It is used to control broadleaf weeds while leaving grasses and conifers unaffected.

HERBICIDES OF HISTORICAL INTEREST

- 2,4,5-Trichlorophenoxyacetic acid (2,4,5-T) was a widely used broadleaf herbicide until being phased out starting in the late 1970s. While 2,4,5-T itself is of only moderate toxicity, the manufacturing process for 2,4,5-T contaminates this chemical with trace amounts of 2,3,7,8-tetrachlorodibenzo-p-dioxin (TCDD). TCDD is extremely toxic to humans. With proper temperature control during production of 2,4,5-T, TCDD levels can be held to about.005 ppm. Before the TCDD risk was well understood, early production facilities lacked proper temperature controls. Individual batches tested later were found to have as much as 60 ppm of TCDD.

- 2,4,5-T was withdrawn from use in the USA in 1983, at a time of heightened public sensitivity about chemical hazards in the environment. Public concern about dioxins was high, and production and use of other (non-herbicide) chemicals potentially containing TCDD contamination was also withdrawn. These included pentachlorophenol (a wood preservative) and PCBs (mainly used as stabilizing agents in transformer oil). Some feel that the 2,4,5-T withdrawal was not based on sound science. 2,4,5-T has since largely been replaced by dicamba and triclopyr.
- Agent Orange was a herbicide blend used by the U.S. military in Vietnam between January 1965 and April 1970 as a defoliant. It was a 50/50 mixture of the *n*-butyl esters of 2,4,5-T and 2,4-D. Because of TCDD contamination in the 2,4,5-T component, it has been blamed for serious illnesses in many veterans and Vietnamese people who were exposed to it.
- However, research on populations exposed to its dioxin contaminant have been inconsistent and inconclusive. Agent Orange often had much higher levels of TCDD than 2,4,5-T used in the US. The name *Agent Orange* is derived from the orange colour-coded stripe used by the Army on barrels containing the product. It is worth noting that there were other blends of synthetic auxins at the time of the Vietnam War whose containers were recognized by their colours, such as Agent Purple and Agent Pink.

FUNGICIDE

Fungicides are chemical compounds used to prevent the spread of fungi or plants in gardens and crops, which can cause serious damage resulting in loss of yield and thus profit. Though oomycetes are not fungi, they use the same mechanisms to infect plants and therefore in phytopathology chemicals used to control oomycetes are also referred to as fungicides. Fungicides are also used to fight fungal infections.

Fungicides can either be contact or systemic. A contact fungicide kills fungi when sprayed on its surface; a systemic fungicide has to be absorbed by the plant. The majority of fungicides that can be bought retail are sold in a liquid form. The most common active ingrediant is sulfur, running at 0.08% for the weaker concentrates, and has high as.5% for the more potent fungicides. In powdered form, the concentration is usually around 90%, and is very toxic.

Other active ingrediants in different brands include neem oil, rosemary oil, jojoba oil, and the bacterium *Bacillus subtilis*. Fungicide residues have been found on food for human consumption, mostly from post-harvest treatments. Some fungicides are dangerous to human health, such as Vinclozolin, which has now been removed from use.

FUNGICIDE RESISTANCE

Pathogens respond to the use of fungicides by evolving resistance. In the field several mechanisms of resistance have been identified. The evolution of fungicide resistance can be gradual or sudden. In qualitative or discrete resistance a mutation (normally to a single gene) produces a race of a fungus with a high degree of resistance. Such resistant varieties also tend to show stability, persisting after the fungicide has been removed from the market. For example sugar beet leaf blotch remains resistant to azoles years after they were no longer used for control of the disease. This is because such mutations often have a high selection pressure when the fungicide is used, but there is low selection pressure to remove them in the absence of the fungicide.

In instances where resistance occurs more gradually a

shift in sensitivity in the pathogen to the fungicide can be seen. Such resistance is polygenic – an accumulation of many mutation in different genes each having a small additive effect. This type of resistance is known as quantitative or continuous resistance. In this kind of resistance the pathogen population will revert back to a sensitive state if the fungicide is no longer applied.

Little is known about how variations in fungicide treatment affect the selection pressure to evolve resistance to that fungicide. Evidence shows that the doses that provide the most control of the disease also provide the largest selection pressure to acquire resistance, and that lower doses decreased the selection pressure.

In some cases when a pathogen evolves resistance to one fungicide it automatically obtains resistance to others – a phenomenon known as cross resistance. These additional fungicides are normally of the same chemical family or have the same mode of action, or can be detoxified by the same mechanism. Sometimes negative cross resistance occurs, where resistance to one chemical class of fungicides leads to an increase in sensitivity to a different chemical class of fungicides. This has been seen with carbendazim and diethofencarb.

There are also recorded incidences of pathogens evolving multiple drug resistance – resistance to two chemically different fungicides by separate mutation events. For example *Botrytis cinerea* is resistant to both azoles and dicarboximide fungicides.

There are several routes by which pathogens can evolve fungicide resistance. The most common mechanism appears to be alternation of the target site, particular as a defence against single site of action fungicides. For example Black Sigatoka, an economically important pathogen of banana, is resistant to the QoI fungicides, due to a single nucleotide change resulting one amino acid (glycine) being replaced by another (alanine) in the target protein of the QoI fungicides, cytochrome b. This presumably disrupts the binding of the fungicide to the protein, rendering the fungicide ineffective.

Upregulation of target genes can also render the fungicide ineffective. This is seen in DMI resistant strains of *Venturia inaequalis*. Resistance to fungicides can also be developed by efficient efflux of the fungicide out of the cell. *Septoria tritici* has developed multiple drug resistance using this mechanism. The pathogen had 5 ABC type transporters with overlapping substrate specificities that together work to effectively pump toxic chemicals out of the cell. Fungi may also develop metabolic pathways that circumvent the target protein, or acquire enzymes that enable metabolism of the fungicide to a harmless substance.

Index

A

B

C

Q

R

S

T

U

V

X